SPACE SCIENCE IN THE TWENTY-FIRST CENTURY: IMPERATIVES FOR THE DECADES 1995 TO 2015

SOLAR AND SPACE PHYSICS

Task Group on Solar and Space Physics
Space Science Board
Commission on Physical Sciences, Mathematics, and Resources
National Research Council

NATIONAL ACADEMY PRESS
Washington, D.C. 1988

National Academy Press • 2101 Constitution Avenue, N.W. • Washington, D. C. 20418

NOTICE: The project that is the subject of this report was approved by the Governing Board of the National Research Council, whose members are drawn from the councils of the National Academy of Sciences, the National Academy of Engineering, and the Institute of Medicine. The members of the committee responsible for the report were chosen for their special competences and with regard for appropriate balance.

This report has been reviewed by a group other than the authors according to procedures approved by a Report Review Committee consisting of members of the National Academy of Sciences, the National Academy of Engineering, and the Institute of Medicine.

The National Academy of Sciences is a private, nonprofit, self-perpetuating society of distinguished scholars engaged in scientific and engineering research, dedicated to the furtherance of science and technology and to their use for the general welfare. Upon the authority of the charter granted to it by the Congress in 1863, the Academy has a mandate that requires it to advise the federal government on scientific and technical matters. Dr. Frank Press is president of the National Academy of Sciences.

The National Academy of Engineering was established in 1964, under the charter of the National Academy of Sciences, as a parallel organization of outstanding engineers. It is autonomous in its administration and in the selection of its members, sharing with the National Academy of Sciences the responsibility for advising the federal government. The National Academy of Engineering also sponsors engineering programs aimed at meeting national needs, encourages education and research, and recognizes the superior achievements of engineers. Dr. Robert M. White is president of the National Academy of Engineering.

The Institute of Medicine was established in 1970 by the National Academy of Sciences to secure the services of eminent members of appropriate professions in the examination of policy matters pertaining to the health of the public. The Institute acts under the responsibility given to the National Academy of Sciences by its congressional charter to be an adviser to the federal government and, upon its own initiative, to identify issues of medical care, research, and education. Dr. Samuel O. Thier is president of the Institute of Medicine.

The National Research Council was organized by the National Academy of Sciences in 1916 to associate the broad community of science and technology with the Academy's purposes of furthering knowledge and advising the federal government. Functioning in accordance with general policies determined by the Academy, the Council has become the principal operating agency of both the National Academy of Sciences and the National Academy of Engineering in providing services to the government, the public, and the scientific and engineering communities. The Council is administered jointly by both Academies and the Institute of Medicine. Dr. Frank Press and Dr. Robert M. White are chairman and vice chairman, respectively, of the National Research Council.

Support for this project was provided by Contract NASW 3482 between the National Academy of Sciences and the National Aeronautics and Space Administration.

Library of Congress Catalog Card Number 87-43332

ISBN 0-309-03848-0

Printed in the United States of America

TASK GROUP ON SOLAR AND SPACE PHYSICS

Frederick Scarf, TRW, Chairman
Roger M. Bonnet, Agence Spatiale Europeene
Guenter E. Brueckner, Naval Research Laboratory
Alexander Dessler, Marshall Space Flight Center
William B. Hanson, University of Texas at Dallas
Thomas Holzer, National Center for Atmospheric Research
Francis S. Johnson, University of Texas at Dallas
Stamatios Krimigis, Applied Physics Laboratory
Louis Lanzerotti, Bell Laboratories
John Leibacher, National Solar Observatory
Robert MacQueen, National Center for Atmospheric Research
Carl E. McIlwain, University of California, San Diego
Andrew Nagy, University of Michigan
Eugene N. Parker, University of Chicago
George Paulikas, Aerospace Corporation
Raymond G. Roble, National Center for Atmospheric Research
Christopher Russell, University of California at Los Angeles
James Van Allen, University of Iowa

Richard C. Hart, *Staff Officer*
Carmela J. Chamberlain, *Secretary*

STEERING GROUP

Thomas M. Donahue, University of Michigan, Chairman
Don L. Anderson, California Institute of Technology
D. James Baker, Joint Oceanographic Institutions, Inc.
Robert W. Berliner, Pew Scholars Program, Yale University
Bernard F. Burke, Massachusetts Institute of Technology
A. G. W. Cameron, Harvard College Observatory
George B. Field, Center for Astrophysics, Harvard University
Herbert Friedman, Naval Research Laboratory
Donald M. Hunten, University of Arizona
Francis S. Johnson, University of Texas at Dallas
Robert Kretsinger, University of Virginia
Stamatios M. Krimigis, Applied Physics Laboratory
Eugene H. Levy, University of Arizona
Frank B. McDonald, NASA Headquarters
John E. Naugle, Chevy Chase, Maryland
Joseph M. Reynolds, The Louisiana State University
Frederick L. Scarf, TRW Systems Park
Scott N. Swisher, Michigan State University
David A. Usher, Cornell University
James A. Van Allen, University of Iowa
Rainer Weiss, Massachusetts Institute of Technology

Dean P. Kastel, *Study Director*
Ceres M. Rangos, *Secretary*

SPACE SCIENCE BOARD

COMMISSION ON PHYSICAL SCIENCES, MATHEMATICS, AND RESOURCES

Foreword

Early in 1984, NASA asked the Space Science Board to undertake a study to determine the principal scientific issues that the disciplines of space science would face during the period from about 1995 to 2015. This request was made partly because NASA expected the Space Station to become available at the beginning of this period, and partly because the missions needed to implement research strategies previously developed by the various committees of the board should have been launched or their development under way by that time. A two-year study was called for. To carry out the study the board put together task groups in earth sciences, planetary and lunar exploration, solar system space physics, astronomy and astrophysics, fundamental physics and chemistry (relativistic gravitation and microgravity sciences), and life sciences. Responsibility for the study was vested in a steering group whose members consisted of the task group chairmen plus other senior representatives of the space science disciplines. To the board's good fortune, distinguished scientists from many countries other than the United States participated in this study.

The findings of the study are published in seven volumes: six task group reports, of which this volume is one, and an overview report of the steering group. I commend this and all the other task group reports to the reader for an understanding of the challenges that confront the space sciences and the insights they promise for the next century. The official recommendations of the study are those to be found in the steering group's overview.

Thomas M. Donahue, Chairman
Space Science Board

Contents

1
Introduction

To understand the relationship between natural events on Earth and changes in the Sun has been one of man's enduring intellectual quests. The scientific discipline we now know as solar system space physics is the modern culmination of efforts to comprehend the relationships among a broad range of naturally occurring physical effects including solar phenomena, terrestrial magnetism, and the aurora. Understanding the solutions to these basic physics problems requires the study of ionized gases (plasmas), magnetohydrodynamics, and particle physics.

Space physics as an identifiable discipline began with the launch of the first earth satellites in the late 1950s and the discovery in 1958 of the Van Allen radiation belts. The phenomena associated with this field of study are among the earliest recorded observations in many parts of the world. The ancient Greeks were puzzled by the "fire" in the upper atmosphere that we now call the aurora; there are several possible references to the aurora in ancient Chinese writings before 2000 B.C.; there are also passages in the first chapter of Ezekiel with vivid descriptions of what we now recognize as auroral formations.

The observation of sunspots by Galileo in 1610 led to the eighteenth-century discovery of the 11-year solar sunspot cycle and the recognition that there was a connection between sunspot variability and auroral activity. The large reduction of sunspots during the second half of the seventeenth century during a period of unusually cool weather in Europe suggests a tantalizing connection between some aspects of solar activity and climate.

This possible link between solar activity and terrestrial phenomena could not be studied in detail until this century. We now know that, in addition to the atmosphere that surrounds us, there exists a region, at higher altitudes, consisting of an electrically conducting plasma permeated by the Earth's magnetic field. It is called the "magnetosphere" because its structure and many of its processes are controlled by the magnetic field. Since the early years of the space program, we have learned that the Sun has its own magnetosphere consisting of a hot (million degree Kelvin) magnetized plasma wind (the solar wind) that extends beyond the orbits of the planets and fills interplanetary space, forming a distinct cavity in the nearby interstellar medium—the "heliosphere."

Using knowledge gained over the past 25 years, we can now begin to identify some of the physical mechanisms linking the Sun to our near-Earth environment. For example, motions in the convective layers of the Sun are believed to generate the solar magnetic field and solar wind variations; these in turn affect the Earth's magnetosphere and regulate the amount of plasma energy incident on the Earth's polar caps. Associated magnetospheric activity drives strong winds in the upper atmosphere and may influence the dynamical and chemical composition of the mesosphere and stratosphere as well. The upper atmosphere, in turn, is the major source of heavy ions in the magnetosphere. Further, current research suggests that small percentage changes (about 0.5 percent) in the total energy output of the Sun (the solar "constant") may influence short-term terrestrial climate. These and other speculative suggestions should be addressed as part of a comprehensive research program in solar-terrestrial physics because of their potential importance for the Earth. Indeed, the Earth and its space environment contain coupled phenomena and need to be studied as a system—from the Sun and its plasma environment to the Earth's magnetosphere, atmosphere, oceans, and biota.

Discoveries in solar and space physics over the past 25 years have inspired a number of developments in theoretical plasma physics. Concepts in charged particle transport theory, developed to describe the behavior of energetic particles in the solar wind and magnetosphere, are routinely used in studying extragalactic radio sources and laboratory plasmas. Magnetic field reconnection (involving the explosive conversion of electromagnetic energy into particle energy), collisionless shock waves, electrostatic shocks, and hydromagnetic turbulence are also among the fundamental plasma phenomena first studied and elucidated in analyses of solar and space plasmas.

Subsequently, these and other concepts have found application to related branches of plasma physics, such as nuclear fusion. The development of space plasma physics since the 1960s has influenced nuclear fusion research. Pitch-angle scattering and magnetic reconnection are now tools of laboratory plasma theory, while ideas developed in fusion work have influenced space plasma science in important ways. Thus, the language of plasma physics links two very important scientific endeavors: the search for a limitless supply of clean energy through thermonuclear fusion and the exploration and understanding of our solar system environment, most of which is in the plasma state.

New concepts developed in studies of solar and space plasmas find important applications to astrophysical problems as well as to laboratory plasmas. For example, the structure of collisionless shock waves can be resolved only by spacecraft instruments. Such shocks are invoked in some current models of star formation. Furthermore, the study of propagating interplanetary shocks has contributed to understanding and modeling of acceleration of cosmic rays by shocks. Particle acceleration via direct electric fields, observed in the Earth's magnetosphere, has been invoked in acceleration models of pulsar magnetospheres. The subject of cosmic-ray transport owes much to detailed in situ studies of the solar wind. Some stellar winds are thought to be associated with stars that, like the Sun, have convective outer layers, while winds of more massive stars are driven by radiation pressure. Explanations of physical phenomena in astrophysical objects that will remain forever inaccessible to direct observation rest heavily on insights obtained through studies of solar system plasmas accessible to in situ observations.

Even though space plasma physics is a mature subject, new observations continue to reveal facets of the physics not recognized previously. For example, observations of "spokes" in Saturn's rings seemed to highlight the importance of electromagnetic forces on charged dust particles. Similarly, the interaction of dust and plasma in comets is thought to be a central element in understanding the formation of comet ion tails. Such observations have given rise to the study of "gravito-electrodynamics" in dusty plasmas, which in turn has important applications to the understanding of the formation and evolution of the solar system, as pointed out by Alfven some years ago.

The understanding of the near-Earth space environment is not only a basic research enterprise; it also has extremely important practical aspects. Space is being used increasingly for many different scientific, commercial, and national security purposes. Well-known examples include communications and surveillance satellites and such scientific platforms as the Space Telescope and the Space Station. These space vehicles must function continuously in the near-Earth environment, subject to the dynamic variations of the heliosphere, the magnetosphere, and the upper atmosphere. It is well established that many spacecraft systems and subsystems exhibit anomalies, or even failures, under the influence of magnetospheric substorms, geomagnetic storms, and solar flares. Processes such as spacecraft charging and "single-event upsets" (owing to highly ionizing energetic particles) in processor memories make the day-to-day operation of space systems difficult. Finally, these aspects of the near-Earth environment become particularly important in view of the planned long-term presence of man in space. The complement of programs outlined in this report will allow us to model the global geospace environment and will thus allow us to develop a global predictive capability. This, in turn, should permit substantial improvements in our abilities to operate all space-based systems in the near-Earth region.

We have advanced well beyond the exploratory stages in solar and space physics, with some notable exceptions—the solar interior, the environment near the Sun where the solar wind is accelerated, the atmospheres of some of the planets, and the boundary of the heliosphere. The phenomenological approach appropriate to a young science still in its discovery phase has progressed to a more mature approach where focused and quantitative investigations are made, interactive regimes are studied, and theory and modeling play a central role in advancing understanding.

The future solar and space physics program will require tools and techniques substantially different from those of the past. Continued progress will require development of complex, multifaceted, experimental and observational projects that will be technologically challenging. The task group believes that the anticipated scientific contributions fully justify the proposed undertakings.

The purpose of this report is to develop an overall program of space research that will address the most significant topics in this discipline, that will clearly define the priority of investigations, and that will be affordable by NASA.

Several other National Research Council reports that are related to this document have appeared in recent years. The Colgate report (*Space Plasma Physics: The Study of Solar-System Plasmas*, 1978) reviewed the status of the field and concluded that "space plasma physics is intrinsically an important branch of science." The Kennel report (*Solar-System Space Physics in the 1980s: A Research Strategy*, 1980) laid out the scientific goals and objectives for the field. Other reports (*Solar-Terrestrial Research in the 1980s*, 1981; *National Solar-Terrestrial Research Program*, 1984) integrated the ground-based segment of the field and stated priorities for its implementation. *The Physics of the Sun* (1985) reviewed the scientific content of solar physics and described future research directions. *A Strategy for the Explorer Program for Solar and Space Physics* (1984) emphasized the need for a revitalized Explorer program for solar and space physics and outlined several specific examples of scientific investigations. *An Implementation Plan for Priorities in Solar-System Space Physics* (1985) updated the scientific goals and objectives of solar and space physics research from the Kennel report and developed the prioritized implementation plan for NASA that would accomplish these aims.

Chapter 2 of the task group's report reviews those scientific objectives, and Chapter 3 describes the status of solar and space physics expected in 1995, assuming a number of space programs proceed according to present plans.

In Chapter 4, a variety of new programs intended for implementation after 1995 are identified that will employ new techniques for the investigation of outstanding scientific questions, will address new questions that arise as natural extensions of previous studies, and will allow the pursuit of new topics that we cannot address at present. This chapter also describes the developments in technology that will be required for the era after 1995; Chapter 5 summarizes the technology needs.

This report also contains a number of appendixes—reports (or excerpts of reports) of workshops that were conducted by NASA in support of this study. The workshops brought together scientists from diverse backgrounds to explore new ideas and technologies for space science research. Those efforts were an important part of the study—they were largely responsible for some of the new initiatives proposed in Chapter 4.

2
Scientific Objectives

The following summary of the scientific objectives of solar and space physics is taken from the NRC report *An Implementation Plan for Priorities in Solar and Space Physics* (1985), which is adapted from the Kennel report (*Solar-System Space Physics in the 1980s: A Research Strategy*, 1980) with appropriate changes and updates.

SOLAR PHYSICS

Major advances in our understanding of the Sun were made in the 1970s and the early 1980s (see Figure 2.1a). Most, if not all, of the magnetic flux that emerges from the convective zone is subsequently compressed into small regions of strong (1200 to 2000 G) field, a fact that is still not understood theoretically. Observations confirmed earlier predictions that the 5-min photospheric oscillation, discovered in the early 1960s, is a global phenomenon. This discovery has made "helioseismology" possible, by which the depth of the convective zone and the rotation below the photosphere have been inferred. In addition, by ruling out the classical model of coronal heating by acoustic waves, observations from the ground and from OSO-8 raised anew the question of what maintains the corona's high temperature. Coronal holes were among the major discoveries of the 1970s. White-light, extreme ultraviolet, and x-ray observations suggested how coronal holes are related to the convective zone and to the solar wind. Skylab observations unambiguously identified magnetic arches as the basic structure of coronal flares. This perception altered our theoretical picture of solar flares and clarified the need for a coordinated multiinstrument attack, which was initiated with the Solar Maximum Mission.

To better understand all the processes linking the solar interior to the corona, we need to study (see Figure 2.1b) the following:

- The Sun's global circulation—how it reflects interior dynamics, is linked to luminosity modifications, and is related to the solar cycle.
- The interactions of solar plasma with strong magnetic fields—active regions, sunspots, and fine-scale magnetic knots—and how solar flare energy is released to the heliosphere.
- The energy sources of the solar atmosphere and corona and the physics of the Sun's large-scale weak magnetic field.

PHYSICS OF THE HELIOSPHERE

The Sun is the only stellar exosphere where complex phenomena common to all stars can be studied in situ. Observations of the Sun and the heliosphere (the plasma envelope of the Sun extending from the corona to the interstellar medium) provide the basis for interpreting a variety of phenomena ranging from x-ray and gamma-ray radiation to cyclical activity and long-term evolution. The Sun, together with the heliosphere and planetary magnetospheres and atmospheres, makes up an immense laboratory that exhibits complex magnetohydrodynamical and plasma physical phenomena whose study enhances our understanding of basic physical laws as well as our understanding of the influences of the Sun on our terrestrial environment.

The solar wind has been studied near the Earth since 1961. At the present time, measurements of the solar wind have primarily been extended to within Mercury's orbit (0.3 AU) and past Pluto's orbit but have been confined to near the ecliptic plane. Quantitative models of high-speed solar wind streams and flare-produced shocks have been developed and tested against data obtained near the ecliptic. The realization that high-speed streams originate in

A SOLAR PHYSICS--RECENT ACCOMPLISHMENTS

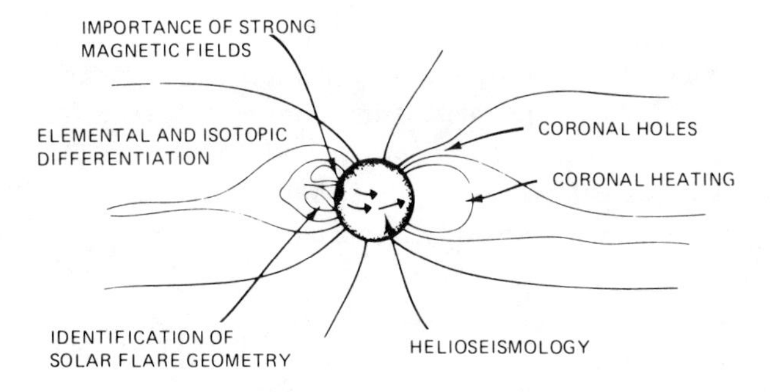

B SOLAR PHYSICS–OBJECTIVES

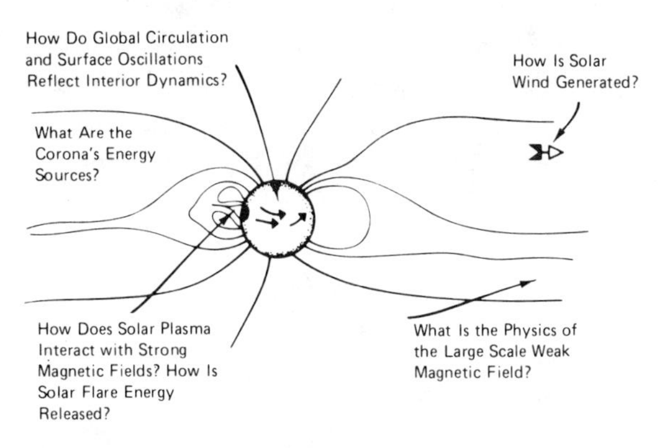

C SOLAR PHYSICS–REQUIRED MEASUREMENTS

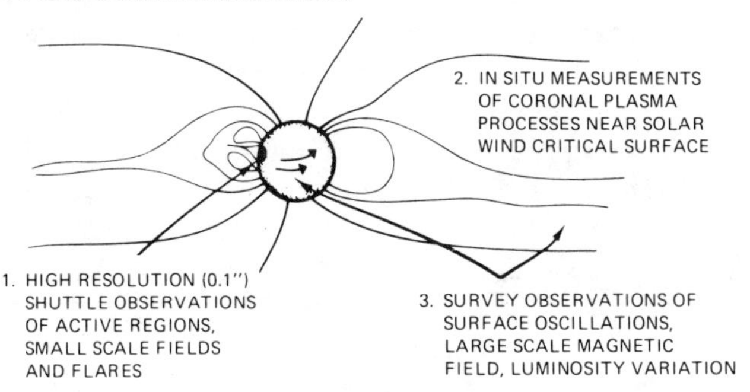

FIGURE 2.1 Solar physics: status, objectives, and recommendations. In this series of sketches of the Sun and its coronal magnetic field some recent accomplishments in solar physics are illustrated. (A) Questions that can be fruitfully attacked in the 1980s and 1990s. (B) The principal research programs needed to answer these questions. (C) The Sun and solar corona within 5 solar radii. The influence of the processes occurring within this region extends throughout interplanetary space via the solar wind.

the rapidly diverging magnetic-flux tubes of coronal holes has reoriented much solar wind research. The sector structure of the solar wind has been unambiguously related to a magnetic neutral sheet of solar system scale that connects to the large-scale magnetic field of the rotating Sun. Finally, microscopic plasma processes have been shown to regulate solar wind thermal conduction and diffusion and, possibly, local acceleration of particles in solar wind structures.

To understand better the transport of energy, momentum, energetic particles, plasma, and magnetic field through interplanetary space, we need to study the following:

- First and foremost, the solar processes that govern the generation, structure, and variability of the solar wind.
- The three-dimensional properties of the solar wind and heliosphere.
- The plasma processes that regulate solar wind transport and accelerate energetic particles throughout the heliosphere.

MAGNETOSPHERIC PHYSICS

New processes regulating Earth's magnetic interactions with the solar wind were discovered in the 1970s and early 1980s (see Figure 2.2a). For example, unsteady plasma flows that apparently originate deep in the geomagnetic tail and deposit their energy in the inner magnetosphere and polar atmosphere were observed. Observations of impulsive energetic particle acceleration suggested that the cross-tail electric field is also highly unsteady. The discovery of energetic ionospheric ions in the near-tail and inner magnetosphere forced a reevaluation of our ideas concerning the origin and circulation of magnetospheric plasma. The 1982-1983 ISEE-3 reconnaissance of the geotail out to 150 R_e provided significant first-order information on tail dynamics during quiescent and substorm conditions.

Our understanding of many individual processes became more quantitative. The coupling of magnetospheric motions and energy fluxes to the thermosphere was observed and modeled. Currents flowing along the Earth's magnetic field and connecting the polar ionosphere to the magnetosphere were found to create strong localized electric fields at high altitudes. These fields may accelerate the electrons responsible for intense terrestrial radio bursts and

A MAGNETOSPHERIC PHYSICS—RECENT ACCOMPLISHMENTS

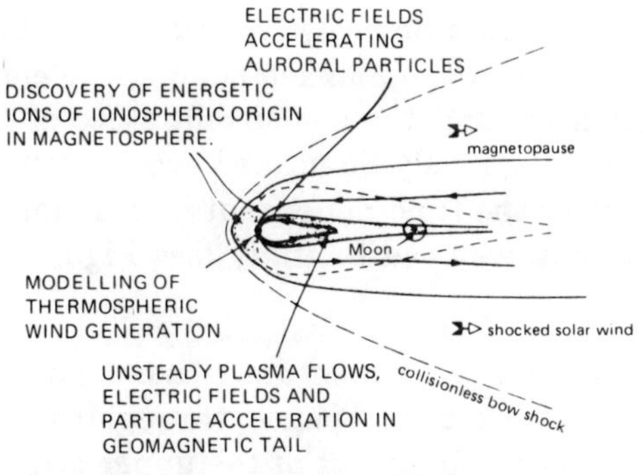

B MAGNETOSPHERIC PHYSICS—OBJECTIVES

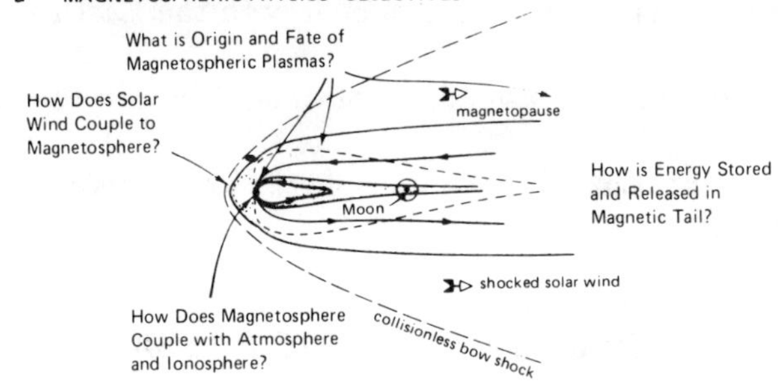

C SIX CRITICAL REGIONS OF MAGNETOSPHERIC PHYSICS

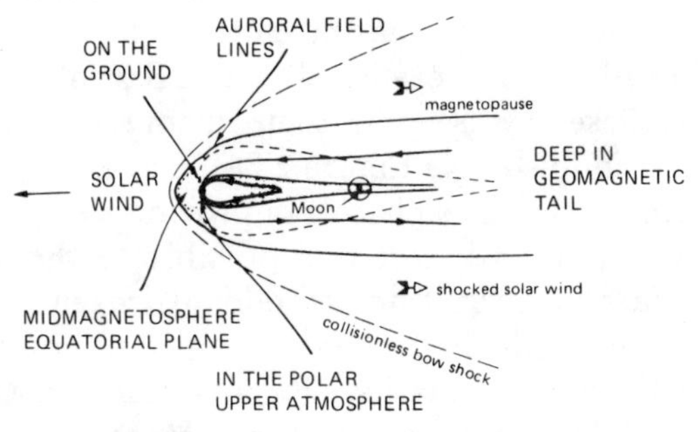

FIGURE 2.2 Magnetospheric physics: status, objectives, and recommendations. Shown here is the Earth's magnetosphere—the cavity formed by the interaction of the solar wind with the Earth's magnetic field. A collisionless bow shock stands upstream of the magnetopause, the boundary separating shocked solar wind from the magnetosphere proper. The Moon is 60 earth radii from the Earth; the Earth's magnetic tail is thought to extend some thousand earth radii downstream. (A) Some recent achievements in magnetospheric physics. (B) Objectives that can motivate research programs in the 1980s and 1990s. (C) The six critical regions where simultaneous studies are needed to help construct a global picture of magnetospheric dynamics.

auroral arcs. Thus, the problem of auroral particle acceleration is nearing quantitative understanding. By contrast, the relationship between energy circulating in the magnetosphere, the energy dissipated in the atmosphere, and the concurrent state of the solar wind has not been unambiguously quantified even today.

To understand better the time-dependent interaction between the solar wind and Earth, we need to study (see Figure 2.2b) the following:

- The transport of energy, momentum, plasma, and magnetic and electric fields across the magnetopause, through the magnetosphere and ionosphere, and into or out of the upper atmosphere.
- The storage and release of energy in the Earth's magnetic tail.
- The origin and fate of the plasma(s) within the magnetosphere.
- How the Earth's magnetosphere, ionosphere, and atmosphere interact.

UPPER ATMOSPHERIC PHYSICS

The upper atmosphere has traditionally been divided into the stratosphere, mesosphere, thermosphere (and ionosphere), and exosphere, in order of increasing altitude. Recent research makes it clear that these layers—and their chemistry, dynamics, and transport—are coupled (see Figure 2.3a). For example, downward transport from the thermosphere can provide a source of nitrogen compounds to the mesosphere and possibly to the upper stratosphere. The catalytic reactions of odd hydrogen, nitrogen, and chlorine compounds destroy ozone, thereby altering the absorption of solar ultraviolet radiation. Results from three Atmospheric Explorers, which largely quantified the photochemistry of the thermosphere and ionosphere, also illustrate the strength of the electrodynamic coupling of the thermosphere to the magnetosphere. Finally, understanding how the upper and lower atmospheres affect each other will be necessary to complete the description of the chain of solar-terrestrial interactions. This will require considerable improvement in our understanding of the chemistry, dynamics, and radiation balance of the mesosphere and stratosphere, as well as of troposphere-stratosphere exchange processes.

To understand better the entire upper atmosphere as one

A UPPER ATMOSPHERIC PHYSICS—RECENT ACCOMPLISHMENTS

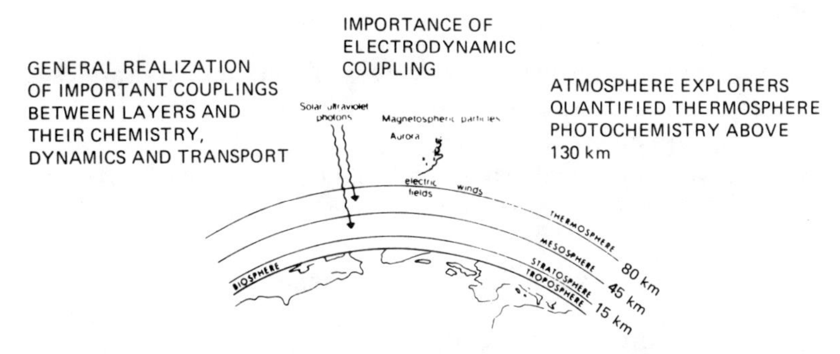

B UPPER ATMOSPHERIC PHYSICS—OBJECTIVES

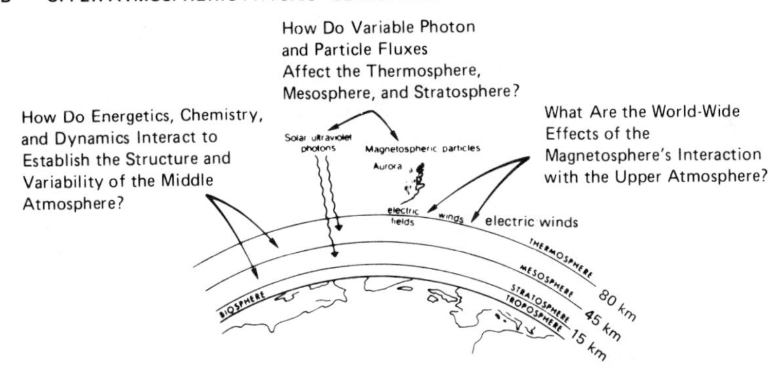

C UPPER ATMOSPHERIC PHYSICS—REQUIRED MEASUREMENTS

A SERIES OF SPACE OBSERVATIONS
mesosphere and stratosphere

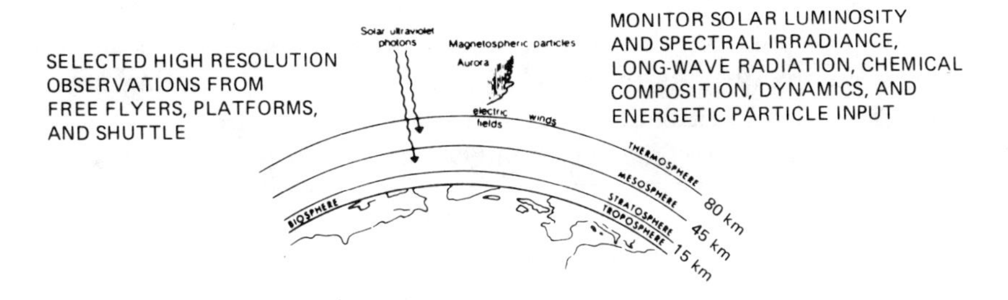

FIGURE 2.3 Upper atmospheric physics: status, objectives, and recommendations. Sketched in these figures are the layers into which the atmosphere has traditionally been divided. Our studies of these layers, and the interacting processes occurring within them, are becoming more integrated. Solar ultraviolet photons deposit their energy largely in the stratosphere and above. The magnetosphere interacts with the upper atmosphere both through energetic plasma deposition and through electric fields, which are generated by magnetospheric motions. Plasma heating and electric fields both couple to upper atmosphere winds.

dynamic, radiating, and chemically active fluid, we should study (see Figure 2.3b) the following:

- The radiant energy balance, chemistry, and dynamics of the mesosphere and stratosphere and their interactions with atmospheric layers above and below.
- The worldwide effects of the magnetosphere's interaction with the polar thermosphere and mesosphere and the role of electric fields in the Earth's atmosphere and space environment.
- The effects of variable photon and energetic particle fluxes on the thermosphere and on chemically active minor constituents of the mesosphere and stratosphere.

SOLAR-TERRESTRIAL COUPLING

Solar-terrestrial coupling is concerned with the interaction of the Sun, the solar wind, and the Earth's magnetosphere, ionosphere, and atmosphere, with particular emphasis on the response of the system to solar variability. For example, a solar flare produces both a strong solar wind shock that initiates a magnetic storm when it passes over the magnetosphere and energetic protons that penetrate deep into the polar atmosphere. Studies of such solar-terrestrial phenomena are of considerable practical importance.

To understand better the effects of the solar cycle, solar activity, and solar wind disturbances upon Earth, we need to do the following:

- Provide, to the extent possible, simultaneous measurements on many links in the chain of interactions coupling solar perturbations to their terrestrial response.
- Create and test increasingly comprehensive quantitative models of these processes.

Whereas 10 years ago it was generally believed that significant effects of solar variability penetrate only as far as the upper atmosphere, some scientists now believe that they also reach the lower atmosphere and so affect weather and climate in ways not yet completely understood. For example, it has recently been suggested that the mean annual temperature in the north temperate zone followed long-term variations of solar activity over the past 70 centuries.

To clarify the possible solar-terrestrial influence on Earth's weather and climate, we need to do the following:

- Determine if variations in solar luminosity and spectral irradiance sufficient to modify weather and climate exist, and understand the solar physics that controls these variations.
- Ascertain whether any processes involving solar and magnetospheric variability can cause measurable changes in the Earth's lower atmosphere.
- Strengthen correlation studies of solar-terrestrial, climatological, and meteorological data.

COMPARATIVE PLANETARY STUDIES

Comparative studies of the interaction of the solar wind with planets and comets highlight the physics pertinent to each and put solar-terrestrial interactions in a broader scientific context. The solar system has a variety of magnetospheres sufficient to make their comparative study fruitful. Because the planets and their satellites have different masses, magnetic fields, rotation periods, surface properties, and atmospheric chemistry, dynamics, and transport, comparative atmospheric and magnetospheric studies can help us to understand these processes in general and possibly to identify terrestrial processes that might otherwise be missed.

In the 1970s, Pioneer and Voyager spacecraft made flyby studies of Jupiter's atmosphere and magnetosphere, the largest and most energetic in the solar system. Pioneer 11 and the Voyagers encountered Saturn in 1979, 1980, and 1981. Mariner 10 flybys discovered an unexpected, highly active magnetosphere at Mercury. Pioneer Venus results suggest that the strong interaction between the solar wind and the upper atmosphere of Venus plays a significant role in the evolution of the atmosphere.

To understand better the interactions of the solar wind with solar system bodies other than Earth, and from their diversity to learn about astrophysical magnetospheres in general, we need to do the following:

- Investigate in situ Mars's solar-wind interaction in order to fill an important gap in comparative magnetospheric studies; previous missions provided little such information.
- Make the first in situ measurements of the plasma, magnetic fields, and neutral gases near a comet.

- Increase our understanding of rapidly rotating magnetospheres involving strong atmospheric and satellite interactions.
- Determine the role of atmospheres in substorms and other magnetospheric processes by orbital studies of Mercury—the only known magnetized planet without a dynamically significant atmosphere.

3
Status Expected in 1995

SOLAR AND HELIOSPHERIC PHYSICS

The subject areas of solar and heliospheric physics encompass a broad range of physical processes; by 1995 we may anticipate continued progress in a number of these areas: the structure and dynamics of the solar interior; variations in the solar luminous output; the emergence and evolution of solar surface magnetic fields, including solar flares; energy and momentum input to the solar corona; global structure and evolution of the heliosphere; and microscopic plasma processes.

Helioseismological studies of the structure and dynamics of the convection zone, the temperature and molecular weight distribution throughout the interior, as well as the radial and latitudinal variation of rotation should be initiated by ground-based networks of instruments as well as instruments on board the solar satellite of the ISTP. However, if our most optimistic expectations of the progress of these projects and their subsequent analysis were justified, we would be only beginning to investigate the possible variation of these properties through the activity cycle.

The specification of the solar interior properties from helioseismological techniques will provide crucial new input into models of the solar dynamo. Indeed, progress in such models—fundamental to our understanding of basic solar cyclic variations—will be stimulated by observational progress. The role of dispersal and diffusion of small-scale magnetic fields in the dynamo process (at least as reflected by surface fields) awaits clues from extended temporal observations of the successor to the SOT (see below).

Solar irradiance will have been monitored on an episodic basis and the solar luminosity will have been monitored on a more continuous basis over more than a complete solar cycle. However, it is unlikely that the solar radius will have been monitored accurately; this, together with the lack of accurate specification of the solar interior properties makes it unlikely that the issues concerning the storage and release of luminous energy (essential in understanding the transport of energy through the solar atmosphere) will have been resolved in a quantitative way.

Extraordinary progress in our view of the fine structure of the solar photosphere and chromosphere will be made by Spacelab 2, Sunlab, and the successor to SOT. It may be anticipated that fundamental new observations of the interaction of solar plasmas and magnetic fields on the spatial and temporal scales at which the basic physical processes occur will have been obtained. To study and understand these processes (for example, changes in magnetic field strength, waves, single pulses, and systematic mass flows), it will be necessary to resolve spatial scales over which significant gradients occur in the local magnetic and nonthermal velocity fields, as well as in the local densities and temperatures. Studies of the plasma-magnetic field interactions on these spatial scales will be directed toward understanding these most fundamental physical processes that can be observed in the solar atmosphere.

The Pinhole Occulter Facility (POF) may be operational by 1995. If so, it will allow hard x-ray observations with a similar spatial resolution as that of SOT. However, no similar capability is anticipated for XUV and EUV spectral observations, or for the crucially important high-energy flare particle signatures. As a result, one may expect limited progress in our understanding of the flare process, which seems to take place also on spatial scales of a few hundred kilometers.

In coronal physics, by 1995, several views of the global proton temperature distribution will have been obtained by new instrumentation, from SPARTAN and SOHO, at least in the region beyond 1.5 solar radii. Inferred coronal outflow speeds—necessarily less reliable—will follow. Thus, our understanding of the physics of the near-solar-wind acceleration will remain largely incomplete, as will our understanding of the origin of the energy flux from the lower solar atmosphere required to drive the wind. Unfortunately, little new information on the three-dimensional structure of the corona will be available by 1995, so all of the long-standing issues concerning the solar longitudinal density distribution of the corona—and also the newer temperature and outflow speed distributions—will remain unresolved. Despite intensive efforts, there will remain a paucity of observations of the initiation of coronal mass ejection phenomena, and suitable diagnostic information on the structure of these events will remain unobserved. Hence, most of these fundamental issues will be unresolved.

Our understanding of the three-dimensional structure of the heliosphere will be improved by the flight of the ISPM, but a second passage of the ISPM will be critical to gaining a reasonable knowledge of latitudinal variations of heliospheric structure during the period of the solar cycle when these variations are expected to be significant. We can also expect some improvement in our knowledge of the latitudinal variation of solar wind mass flux and flow speed (away from solar maximum) from EUV and radio observations that remotely sense the interplanetary medium. Observations of the distant solar wind will be extended to beyond 60 AU, and our present view of distant stream evolution is likely to be confirmed, but it is not clear whether the termination of supersonic solar wind flow will be observed by the Voyager 1 spacecraft, which will have gone some 50 AU in the direction of the solar apex.

Significant progress can be expected in two areas of study involving microscopic plasma processes in the interplanetary medium. Continued theoretical work on the problem of heat conduction and viscosity in a thermally driven stellar wind not dominated by Coulomb collisions should advance to the point where lack of in situ observations of the inner solar wind would prevent further progress. A combined observational and theoretical effort toward understanding the acceleration of particles at interplanetary shocks should meet with some success and provide an improved basis for understanding energetic particle populations in interplanetary space and other astrophysical systems.

We can expect to see some progress in the study of the role of corotating interaction regions on the interplanetary modulation of galactic cosmic rays. There is no reason to believe that we will learn much more about any other modulation mechanisms by 1995.

MAGNETOSPHERIC PHYSICS

The solar system offers a variety of magnetospheres for study. These magnetospheres have sufficiently different internal parameters and boundary conditions to allow testing of the universality of basic magnetospheric concepts. The solar wind interacts with planetary magnetic fields to produce diverse phenomena that involve the storage and energization of charged particles. Also, a small fraction of the energy within a magnetosphere is converted into a broad assortment of radio emissions. Differences and similarities between magnetospheres, if properly understood and formulated, yield physical principles that can be applied throughout the universe.

Sources of Plasma

The solar wind, immediately after its existence was established, was thought to be the source of plasma, particularly energetic plasma, within the Earth's magnetosphere. Nearly two decades later it was discovered that the ionosphere was a source of quantitatively significant flows of plasma into the magnetosphere. Then, consistent with the discovery that the solar wind is a negligible source of plasma for the magnetospheres of Jupiter and Saturn, experimenters have interpreted recent data from the DE satellite as indicating that perhaps most of the Earth's magnetospheric plasma may be supplied by the ionosphere. This question may be partially resolved by AMPTE, which can give a quantitative indication of the ability of solar wind, magnetosheath, and tail plasma to enter the inner magnetosphere, and ISTP, which will make simultaneous, coordinated measurements in the solar wind and the magnetosphere.

Sources of Power

Phenomena within the Earth's magnetosphere are driven by power extracted from the solar wind. The solar wind, as it passes the Earth, drives plasma within the magnetosphere in a circulation pattern that energizes some of the magnetospheric plasma and drives strong magnetically field-aligned currents that connect to the auroral zones. Magnetic merging, in which topologically separate magnetic regions are connected and magnetic energy is released, is still a poorly understood process that is central to most of the discussion of magnetospheric substorms and power delivery to the Earth's magnetosphere and other dynamic processes. It is expected that ISTP will provide data that is necessary to more fully understand this phenomenon. In contrast to the solar wind power source for the Earth's magnetosphere, Jupiter and probably Saturn draw power for their magnetospheres primarily from the kinetic energy of planetary rotation. The flybys of Uranus and Neptune by Voyager will give us two more examples of magnetospheric energization from which to derive general physical principles governing the transfer of power to magnetospheric regions.

Generation of Plasma Waves and Radio Emissions

Cosmic-scale plasma and magnetic field regions are not quiescent. One universal form of activity they exhibit is the generation and amplification of radio waves. Aside from the interest in the radio emissions themselves, they serve as a means of exchanging energy between particles in a low-density plasma where Coulomb collisions are ineffective. Because we can get so close to (even within) the emitting regions of a planetary magnetosphere, we hear all of the symphony of electromagnetic emissions they produce: hiss, chorus, narrow- and broad-band electrostatic waves, and decimetric, decametric, and kilometric radiations. A common theoretical framework has been developed so that when account is taken of differing conditions within the various magnetospheres, radio emissions in one planetary magnetosphere can be related to that in another as different manifestations of the same physical process, although the frequency ranges of the emissions might be quite different.

The Terrestrial Magnetosphere as a System

The International Solar Terrestrial Physics Program will acquire simultaneous measurements throughout key regions in order to understand the behavior of the system as a whole. A coordinated network of spacecraft will permit us to investigate the physical behavior of each key region involved in the analysis of solar-terrestrial plasma physics. WIND will be stationed in the upstream solar wind to observe the interplanetary input function and to observe the escape of energetic magnetosphere particles back into the solar wind. POLAR will (1) observe directly the entry of magnetosheath plasma via the dayside cusp, (2) measure the flow of hot plasma into and out of the ionosphere on auroral field lines, and (3) observe the deposition of particle energy into the ionosphere/atmosphere. CRRES will permit us to observe solar wind-magnetosphere coupling near the magnetic equator, and it will directly measure the interaction of ionosphere and tail plasmas in the ring current and the plasma-sheet storage and transport regions when the apogee is situated over the night hemisphere. GEOTAIL will provide extensive simultaneous measurements of entry, storage, acceleration, and transport in the geomagnetic tail and the tail plasma sheet; and it will also measure plasma entry and transport in magnetopause boundary layers along the dawn and dusk flanks of the magnetosphere. The four CLUSTER spacecraft will provide detailed information on localized current systems and magnetohydrodynamic processes, and SOHO will yield additional information on interplanetary and solar phenomena that can affect the terrestrial magnetosphere.

These ISTP measurements will allow us to develop an understanding of the physical processes occurring in the solar-terrestrial environment. In addition, theoretical studies will provide the framework upon which the empirical understanding from the observations can be both systematized and used to further our basic understanding of other plasma systems.

Solar-Terrestrial Research

In parallel with the major spacecraft efforts addressed above, progress in understanding the solar-terrestrial system as a whole will also occur through the use of existing and planned ground-based research programs in the United States and in nations throughout the world. These include magnetic field and ionospheric arrays established for monitoring and/or campaign purposes and large radar systems for studies of high-altitude plasma convection and transport. During the ISTP program, the international ground arrays will be considerably enhanced and supplemented by more sensitive instruments and, particularly, by considerably advanced data acquisition, storage, and transmission systems. All of these advances will contribute crucially to understanding the global terrestrial magnetosphere system in the context of the ISTP program.

In the United States the ongoing Solar-Terrestrial Theory Program and other theoretical support will be crucial in order to provide the predictive theoretical models for the global system. Input data required for the theory and model advances will occur from both the spacecraft and the ground-based experiments.

UPPER ATMOSPHERE SCIENCE

The upper atmosphere is defined here as the region of the neutral atmosphere above the tropopause (12 km) extending all the way to the exosphere. During the 1970s a great deal of progress was achieved in our understanding of mid- and low-latitude thermospheric photochemistry, mainly as a result of measurements obtained by the Atmosphere Explorer satellites. At the present time the dynamics of the thermosphere is being studied extensively by in situ and remote sensing instruments, which were carried by the Dynamics Explorer 2 satellite, by ground-based optical and radar methods, and by large-scale model studies. The dynamics of the magnetosphere and thermosphere are coupled through field-aligned currents, electric fields, Joule heating, and particle precipitation, which both heats the upper atmosphere and alters the electrical conductivity of the ionosphere. Before the end of this decade, this multipronged study of thermospheric dynamics is expected to lead to significant advances in our understanding of the energy and momentum sources controlling atmospheric motions in these altitude regions. However, gaps will still remain in our understanding of how specific ionospheric and auroral phenomena correspond to magnetospheric sources and/or consequences.

Significant progress in our understanding of the stratosphere, mesosphere, and lower thermosphere (often referred to as the middle atmosphere) is expected as the result of measurements made by the instrument complement of the Upper Atmosphere Research Satellite (UARS), to be launched in the late 1980s. This mission will carry out a comprehensive set of measurements, particularly from the viewpoint of atmospheric chemistry. Global measurements of O_3 and many of the radical species that destroy it (e.g.,

NO_2, ClO) will be acquired simultaneously, allowing for analysis of their interrelationships. Further, the "reservoir" species that sequester these radicals in relatively inert forms will also be measured (e.g., HNO_3, HCl), along with the long lived molecules that are the sources of all these constituents (e.g., N_2O, CH_4). In addition to these chemical measurements, atmospheric winds will be measured simultaneously from UARS, allowing for a more complete analysis of transport processes than ever before possible. An explorer-class mission, MELTER, is also planned for the early 1990s for a focused study of the energetic and dynamic coupling processes between the mesosphere and lower thermosphere. Finally, atmospheric measurements need to be complemented by continued laboratory measurements of the rate coefficients of the important reactions as well as the atomic and molecular parameters of the relevant absorbing and emitting atmospheric species.

These data will be analyzed and interpreted with the aid of multidimensional, chemical-dynamical-radiative models that will include much more complete descriptions of many of the relevant physical and chemical processes and their coupling. The progress in all the various numerical modeling studies that are relevant to atmospheric sciences will be greatly accelerated by advances in computer technology.

A critical aspect of our understanding of the middle atmosphere is the question of what physical and temporal scales are involved in the various photochemical and transport processes. The spatial resolution achievable, currently as well as in the near future, by both satellite observations and numerical models is relatively large and represents a significant barrier to a thorough understanding of many of the smaller scales on which significant processes may well occur.

Another important area of study in the 1990s will continue to be the question of interactions between the middle atmosphere and its neighbors, the troposphere and the thermosphere. Thermospheric coupling includes studies of the particle and solar inputs and corresponding middle atmospheric responses on a global scale. Finally, the important question of the coupling of the troposphere and the middle atmosphere will be a central component of the field as a whole; anthropogenic perturbations to the atmosphere in general and the ozone layer specifically will continue to be an important central theme for middle atmospheric studies.

Global Electric Circuit

The existence of a global electric circuit has been known for many years, and yet we still do not understand the basic processes that drive and control the behavior of this system. According to the classical picture, thunderstorms are the sole generators within the circuit, and, acting together, they maintain a potential difference of about 200 to 600 kV between the highly conducting ionosphere and the surface of the Earth. This potential difference causes a downward conduction current of about 1 to 2 kA from the ionosphere to the ground in the fair-weather dissipative portion of the circuit. The global circuit is believed to be closed through an upward current flow from the Earth's surface, beneath a thunderstorm, to the negative charge at the cloud base. Within a cloud, updrafts and downdrafts and microphysical and electrification processes maintain the charge separation. Although this picture is generally the accepted one today, considerable doubts and uncertainties are associated with many of the macrophysical and microphysical concepts that have been advocated (e.g., are thunderstorms the main generators, and what controls charge separation?). Simple models of the circuit assume that the ionosphere is a highly conductive, equipotential upper boundary; however, in reality, there are significant horizontal potential differences of tens of kilovolts generated by both the ionospheric neutral wind and the solar wind/magnetosphere dynamos. The details of the telluric currents flowing in both the solid earth and oceans are complex and require comprehensive experimental and theoretical investigations.

The problems associated with the global electric circuit cut across numerous disciplines, from magnetospheric convective processes at one end to soil and ocean conductivity issues at the other end.

Some of the more "relevant" ways in which atmospheric electricity may play a potentially important role are as follows:

- Causing interference in man-made systems such as communication cables, power lines, and pipe lines.
- Influencing the spatial distribution and effectiveness of condensation nuclei in the atmosphere.
- Acting as a possible mechanism for the generation of odd nitrogen compounds through lightning processes.

PLANETARY SPACE PHYSICS

There are basically four different types of interactions of plasmas with bodies in the solar system. The plasmas can collide with solid bodies directly. For example, the solar wind strikes the lunar surface and is absorbed. Energetic particles are similarly absorbed by the jovian moons, albeit with sputtering of material that in turn supplies material to the jovian magnetosphere. The rings of Saturn are also subject to such processes.

A second interaction is the deflection of a flowing plasma by a planetary magnetic field. The smallest scale on which such deflection is known to occur is the deflection of solar wind above the lunar terminators when magnetized regions of the lunar surface occur at the lunar limbs. Such deflection occurs on a global scale at Mercury, the Earth, Jupiter, and Saturn and perhaps other planets as well.

If a body has no strong magnetic field but does have an atmosphere, two other interactions can occur, whose effects are sometimes difficult to separate. First, if an ionosphere can form due to strong ionization of the neutral atmosphere, a cold plasma region may form whose pressure is sufficient to exclude the external nonplanetary plasma. This process occurs at Venus, where the gravitationally bound ionosphere usually has sufficient pressure to stand off the solar wind. Fresh comets, strongly outgassing near the Sun, are also thought to have ionospheres. If the ionospheric pressure is insufficient to stand off the solar wind, it is essentially pushed back toward lower altitudes where the atmospheric density is greater. The resulting ion-neutral coupling due to the relative drift of the ions and neutrals transmits pressure to the plasma and aids in the support of the ionosphere. At comets, the outflow of neutral gas assists in this pressure balance. At the top of the ionosphere, the ionopause, a tangential discontinuity, separates the external plasma, the magnetosheath in the case of Venus, from the ionospheric plasma.

The second process that occurs is the direct interaction of the neutral gas and the external plasma without the intermediary of an ionosphere. The neutral atmosphere at a planet such as Venus extends well above the ionopause. This gas can be photoionized by solar extreme ultraviolet or charge exchange with the magnetosheath or solar wind plasma. At a satellite such as Io, the neutral gas can be ionized also by impact ionization. This new source of plasma mass loads the external plasma, slowing it down if it is flowing. Charge exchange can also lead to the creation of fast neutrals that remove momentum from the plasma. This process is thought to create the long tails seen behind Venus and comets and cause the field-aligned currents and distorted magnetic fields seen at Io and Titan.

Finally, it is noted that the solar wind flows much more rapidly than the speed of compressional waves in the solar wind plasma. Hence, when it interacts with a conducting planetary object, shock waves form to deflect the flow around the object. These collisionless shocks are very interesting objects that have been intensively studied at the Earth. However, planetary studies have much to add to this investigation because the properties of the plasmas in the solar system change greatly with heliocentric distance.

Interaction with Unmagnetized, Atmosphereless Bodies

The interaction of the solar wind with the Moon is understood only on the most elementary level. Basically we know that the solar wind is absorbed by the forward hemisphere of the Moon, leaving a wake behind. Because of the limited plasma instrumentation on Explorer 35 and the Apollo subsatellites, we know little about the closure of plasma behind the Moon. Some work has been done with plasma instruments on the lunar surface and much empirical understanding obtained.

The interaction of the radiation belt of Jupiter with its satellites is understood on an elementary level, although much more needs to be done on the role of sputtering in providing a source of jovian plasma. Galileo should contribute greatly in this area. We also have an elementary understanding of the interaction of Saturn's radiation belts with its satellites and rings. We expect new progress in this area through laboratory investigations, theory, and the Saturn orbiter of the Cassini mission. Studies of the interaction of the solar wind with cometary dust are also important. Here, we expect some near-term progress through laboratory data and theoretical treatment.

Deflection by Planetary Magnetospheres

We know little about the solar wind interaction with Mercury except that Mercury has a magnetosphere, bow shock, and transient energetic particle population, suggesting that substorm-like processes occur. Mercury's magnetosphere is a very important one because of its lack of a dynamically important atmosphere. Thus field-aligned currents should not play a dominant role in the magnetosphere of Mercury.

Our studies of the jovian magnetosphere have proceeded from the exploratory phase to the beginnings of intensive investigation. We expect further refinements from the Galileo mission. The jovian magnetosphere is important because of its rapid rotation and the large mass loading by Io deep in the interior of the magnetosphere.

Saturn complements the jovian studies with its extensive ring system that both absorbs and supplies charged particles. Titan is also important as discussed below. We expect little change in our understanding of the saturnian magnetosphere until the data from the Cassini mission are obtained.

Voyager has provided existence data on the magnetosphere of Uranus, leaving us with elementary concepts on the physics of these systems. Similar data will be available from Neptune in 1989.

Plasma-Atmosphere Interactions

The solar wind interaction with Venus is now understood to first order. The present and future data base from Pioneer Venus plus computer modeling should give us sufficient insight to ask more fundamental questions, but large gaps remain in our observational knowledge.

The magnitude of the Mars intrinsic magnetic moment is not known. This will be addressed with the Mars Observer in the early 1990s. However, we will still be ignorant of all the basic plasma processes on Mars, as well as of the processes in the upper neutral atmosphere.

The interaction of the jovian plasma with Io and the interaction of the saturnian plasma with Titan are also key elements in our study of plasma/neutral gas interactions. The former will be addressed briefly by Galileo in one flyby. This may not be sufficient. The latter will be addressed by the Cassini mission to Saturn.

Interest in the interaction of the solar wind with comets is undergoing a strong resurgence owing to the high interest in the Giacobini-Zinner and Halley missions. Theoretical progress should be made by the end of the decade, and there should be some confirmatory data.

SUMMARY

Tables 3.1 and 3.2 provide a graphic summary of the expected status of research by 1995 on the science objectives defined in Chapter 2. Table 3.1 depicts in situ investigations; Table 3.2 depicts remote sensing investigations.

TABLE 3.1 Levels of In Situ Spacecraft Investigation

Discipline	Awaiting Recon- naissance	Reconnais- sance	Exploration	Intensive Study	Physical Under- standing
Solar physics					
Sun, solar corona	Solar Probe				
Heliospheric physics					
Generation of solar wind	Solar Probe				
High-latitude solar wind	ISPM				
In-ecliptic solar wind beyond Saturn					
In-ecliptic solar wind between Mercury and Jupiter					
Heliopause, interstellar medium	•				
Terrestrial, magnetospheric physics					
Magnetosphere <60 R_e Earth magnetic tail, wake			ISTP		
Terrestrial, atmospheric, ionospheric physics					
Thermosphere, ionosphere >150 km					
Mesosphere	•				

TABLE 3.1 (continued)

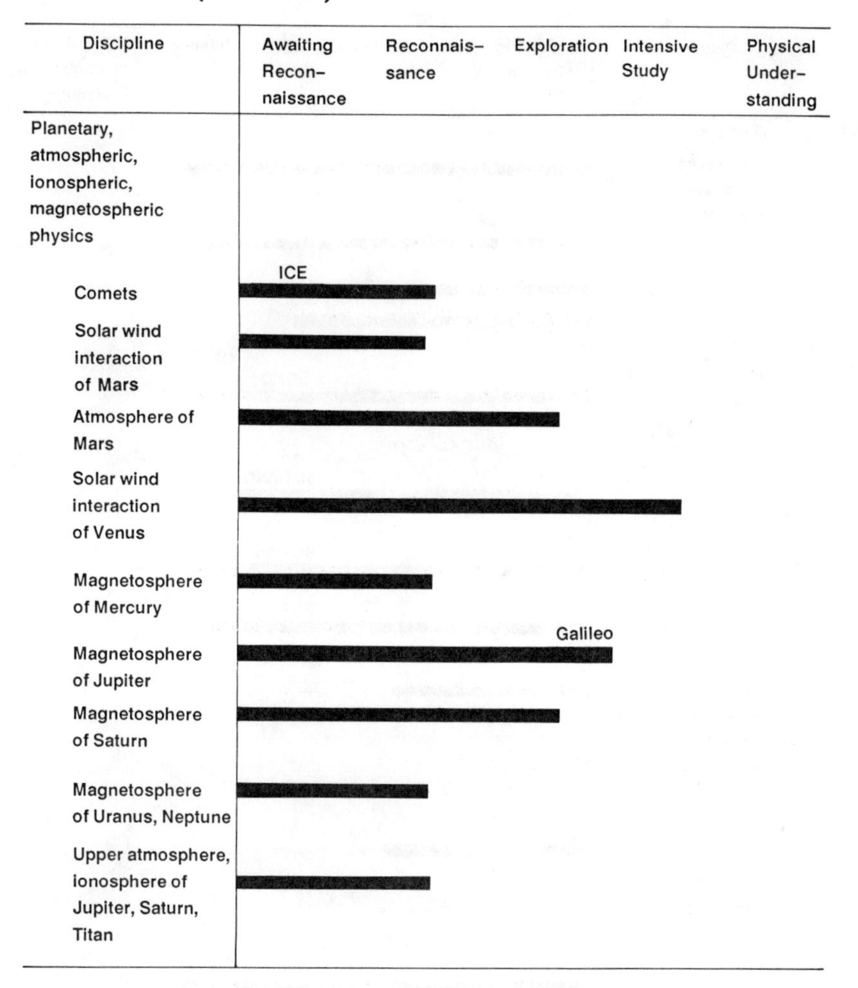

TABLE 3.2 Levels of Remote Sensing Investigation

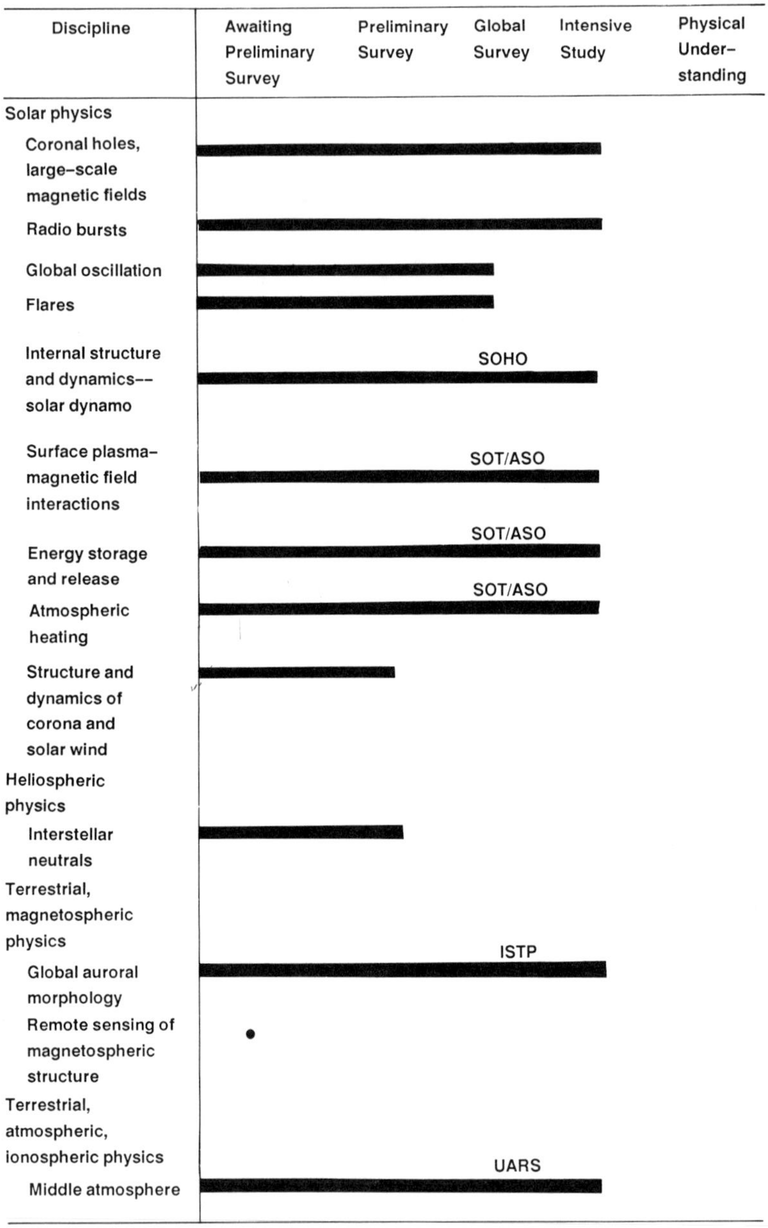

Discipline	Awaiting Preliminary Survey	Preliminary Survey	Global Survey	Intensive Study	Physical Understanding
Solar physics					
Coronal holes, large-scale magnetic fields					
Radio bursts					
Global oscillation					
Flares					
Internal structure and dynamics--solar dynamo			SOHO		
Surface plasma-magnetic field interactions			SOT/ASO		
Energy storage and release			SOT/ASO		
Atmospheric heating			SOT/ASO		
Structure and dynamics of corona and solar wind					
Heliospheric physics					
Interstellar neutrals					
Terrestrial, magnetospheric physics					
Global auroral morphology			ISTP		
Remote sensing of magnetospheric structure	•				
Terrestrial, atmospheric, ionospheric physics					
Middle atmosphere			UARS		

4
New Initiatives: 1995 to 2015

Many of the most novel and exciting initiatives in solar and space plasma physics for the period 1995 to 2015 will involve combining remote sensing or imaging techniques with in situ observations for the study of the Earth's magnetosphere and the analysis of the solar corona, as well as other solar phenomena. These new initiatives fall into four groups, corresponding to the following subdisciplines of solar and space physics: heliospheric physics, terrestrial magnetospheric physics, terrestrial atmospheric physics, and planetary science.

SOLAR AND HELIOSPHERIC PHYSICS

Local Measurements in the Solar Atmosphere

At present, our understanding of the origin of the solar wind is based entirely on theory and remote sensing. Direct measurements of the solar wind plasma, the interplanetary magnetic field, the energetic particle population, and associated wave-particle interactions are available, but only at distances greater than the 0.3 AU perihelion distances of Helios 1 and 2. The task group recommends a Solar Probe mission whose primary objective is to carry out the first in situ observations of the solar wind plasma and fields (electric and magnetic) near the source of the wind in the

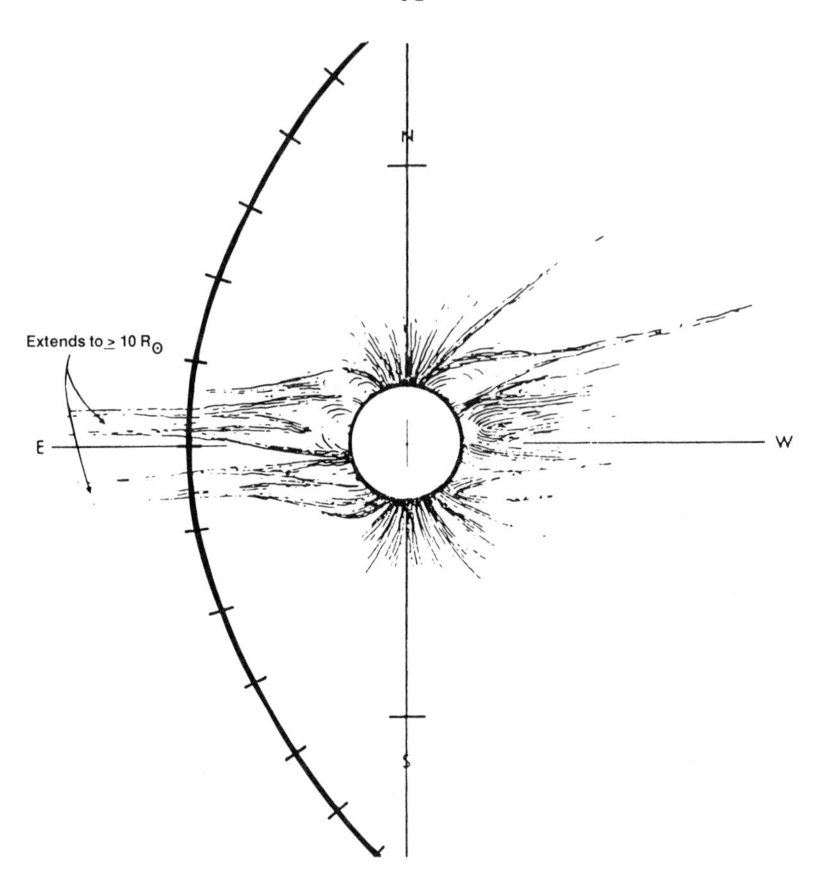

FIGURE 4.1 A trajectory for the Solar Probe.

solar atmosphere. Included will be a detailed study of energetic particles, which will yield important diagnostic data on particle acceleration processes and coronal structure.

The spacecraft must be placed in an orbit that will bring it as close to the Sun as possible and still survive to provide useful data near closest approach. A perihelion distance of 4 solar radii is anticipated, with a local wind speed of about 50 km/s, electron and ion plasma temperatures of about 10^6K, and plasma density and magnetic field strength of less than 10^6 electrons/cm^3 and 10^5 gamma, respectively. A drawing of the Solar Probe trajectory is shown in Figure 4.1.

Theories of solar wind origin place the transition region from subsonic plasma flow to supersonic flow somewhere between 1 and

10 solar radii. Radio scattering experiments on Viking during superior conjunction suggest a critical point closer to 10 solar radii. In situ measurements should clarify this issue.

The location of the critical point and the plasma properties (speed and temperature) of the supersonic wind will depend greatly on the physical processes that heat the corona. Theoretical studies suggest that the proton temperature profile is very sensitive to these heating processes. It is not clear whether the corona contains an extended region of heating (out to as far as 20 solar radii) or undergoes adiabatic expansion beyond the solar surface. Plasma temperature data and observations of the wave types and amplitudes should lead to the identification of the important heating and acceleration mechanisms.

Many other important problems can be studied with Solar Probe, including a detailed characterization of coronal streamers, the place of origin and the boundaries of high-speed and low-speed flows close to the Sun, the extent of heavy element fractionation and elemental abundance variations, and the scale sizes of inhomogeneities and the development of the magnetohydrodynamic turbulence that characterizes the solar wind near 1 AU and beyond. The Solar Probe mission can also study the solar spin down rate through measurements of solar wind angular momentum flux.

Further study needs to be carried out to determine the best method of designing detectors that are required to look in the direction of the Sun.

In the original study, it was assumed that the spacecraft would go to Jupiter, where a gravity assist would send it on course to the inner corona. Our task group learned of a possible alternate trajectory involving a hypersonic flyby in the upper atmosphere of Venus; the two possibilities are sketched in Figure 4.2. It should be possible to add low-thrust propulsion in order to attain an ecliptic orbit around the Sun with a 1-year periodicity so that the probe enters the vicinity of the Sun several times. As shown in Figure 4.2, the Venus flyby technique also yields a very short orbital period.

High-Latitude Solar Studies

The heliosphere is known to have a complicated three-dimensional structure. The magnetic field is a tight spiral near the solar equatorial plane, but is expected to be essentially radial over the

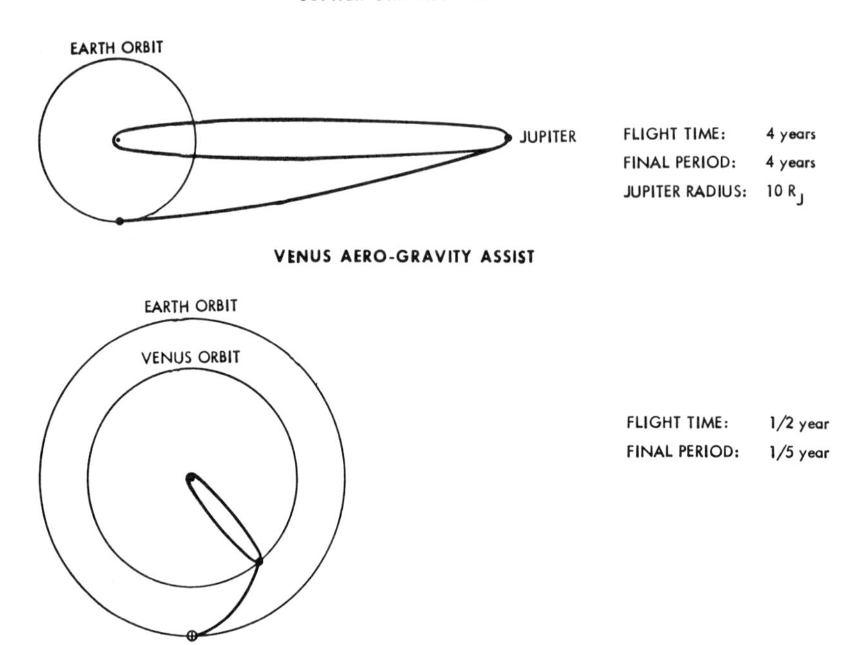

FIGURE 4.2 Two concepts for the trajectory of a Solar Probe mission.

solar poles. Coronal holes, one of the sources of high-speed solar wind, are expected to produce quasi-steady high-speed flows over the solar poles during much of the solar cycle, whereas at low latitudes interacting high- and low-speed flows predominate.

To understand heliospheric conditions at low solar latitudes has required numerous missions, e.g., Explorers, Pioneers, Mariners, and Voyagers. To understand heliospheric conditions at high latitudes will similarly require repeated missions. NASA and ESA will fly the first exploratory mission over the solar poles (Ulysses). However, as with most exploratory missions, Ulysses will probably uncover more questions than it will answer, and follow-on missions will be required.

The objective of the Solar Polar Orbiter (SPO) would be to provide a detailed, repeated study of conditions at all heliographic latitudes. In circular orbit, SPO will observe the heliosphere at constant radius and thus will distinguish latitude from radial effects. With a circular orbit at less than or equal to 1 AU, and thus

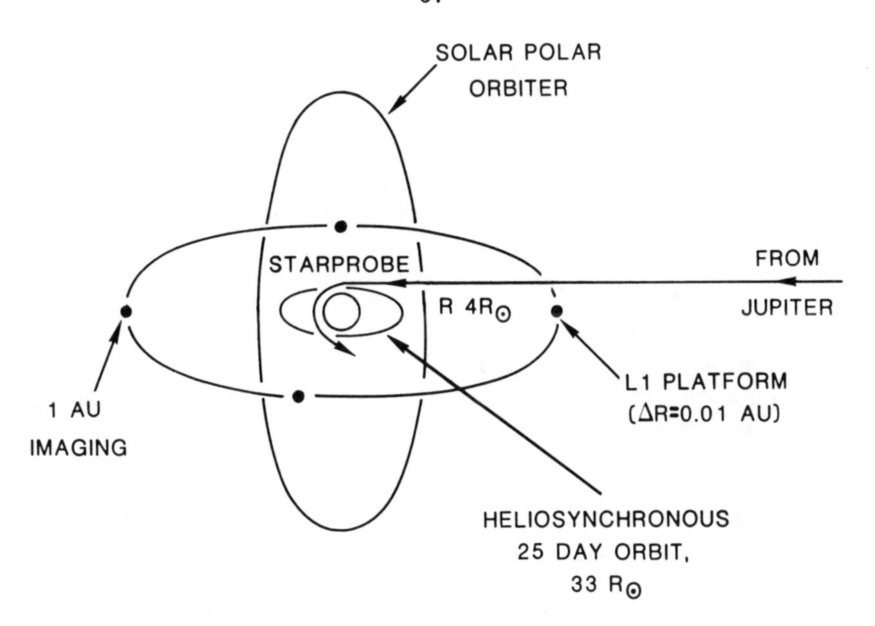

FIGURE 4.3 The orbit of the Solar Polar Orbiter shown together with locations of other solar measurement platforms (Starprobe, the 1-AU observing network, and the Heliosynchronous Orbiter).

an orbital period less than or equal to 1 year, SPO should be able to make several passes over the solar poles in a nominal mission lifetime, and thus distinguish spatial from temporal effects.

No detailed study of an SPO mission has yet been done. However, the required orbit should be achievable through the use of a low-thrust, continuous acceleration propulsion system such as solar-electric propulsion with a final orbit as shown in Figure 4.3.

The SPO spacecraft should carry a full complement of plasma, energetic particle, magnetic field, and radio wave instruments, similar to what is to be flown on ISPM. In addition, SPO should have pointing capability, through the use of a despun platform on a spinning spacecraft, or as a three-axis stabilized spacecraft, for detailed solar observations using a coronagraph, x-ray telescope, and similar photon observing instruments.

The principal technical development required for SPO is a solar-electric propulsion system, or its equivalent, for low-thrust, continuous acceleration. In the cost projections for SPO it is assumed that such development will not be charged against the mission costs, because the need is common to several proposed programs. Also, studies need to be conducted on the impact of a continuous propulsion system on particle, field, and photon instrumentation, and on the measurements these instruments make.

Outer Coronal Physics

The Heliosynchronous Orbiter (HSO), as described in the ESA document *Horizon 2000*, is an instrumented probe orbiting the Sun at about 30 solar radii with a 25-day period, synchronous with the rotation of the Sun (see Figure 4.3).

This mission will be able to address a very broad range of scientific objectives from solar physics, physics of the interplanetary medium, and high-energy astrophysics to relativity:

- Investigation of the morphology and dynamical development of all solar structures from the photosphere to the outer corona from a vantage point close to the Sun (0.15 AU) over a large range of solar latitudes, with frequent access to the solar polar regions. Stereoscopic viewing of structures through motion of the spacecraft. The understanding of the relationship between the thermal structure and heating of the solar corona will ultimately permit the identification of the physical nature of the solar wind acceleration. Imaging of the coronal structures could be achieved by observations at 1 AU.
- Investigation of the three-dimensional structure of the inner heliosphere near or even outside the region where the wind is accelerated.
- Measurements of solar wind particle fields and waves; studies of the heating and acceleration of the solar wind (thermally or wave-driven wind?) with the advantage of a wide latitude coverage.
- Studies of the propagation, acceleration, and modulation of solar energetic particles including the significant reduction of propagation effects with respect to 1 AU. Study of shock wave acceleration.
- Radio sounding of the solar corona as the spacecraft passes behind the Sun.
- Correlative studies of expanding and traveling solar structures and their manifestation in interplanetary space.
- Investigation of the three-dimensional distribution of mass and velocity of interplanetary dust in the inner heliosphere.

- Investigations of the Hermean magnetosphere and remote sensing of Mercury during flybys early in the mission (the last in situ measurements date from Mariner 10 in 1974-1975).
- Establishment of a reference observatory for other missions in the heliosphere, in particular for solar optical remote sensing missions near 1 AU.
- Baseline observations of galactic gamma-ray bursts.
- Performance of relativity experiments (if possible), e.g., determination of J_2 (the second gravitational moment) in the case of a highly elliptic orbit; frame dragging experiments.

The technological problems (propulsion, thermal design, data transmission) pose a considerable challenge. The mission concept is certainly not only attractive to a large scientific community, but it would also be appealing to the general public and from the technological point of view.

1-AU Observing Network

The Sun is the only star that we can observe from different directions, i.e., from any position within the heliosphere. This provides a stereoscopic view of structures whose geometry and energy content cannot be determined because they are either optically thin or because parts of them are not entirely visible from one single viewing condition. In addition, simultaneous observations at different positions inside the heliosphere provide three-dimensional snapshots of the magnetic field and the solar wind, important observations that will give new insight into the mechanisms that govern the wind generation, acceleration, and propagation. Similarly, simultaneous measurements of the irradiance with a set of several spacecraft would allow us to infer what mechanisms induce variations in the solar constant, whether they are due to sunspot luminosity deficiencies compensated by equivalent increases on the hidden solar hemisphere or whether they are in phase over the whole surface and due to global variation of the solar volume. It should also be noted that a 360° network for ecliptic monitoring of flare events might become an indispensable element in any manned mission to another planet.

A set of 4 1-AU spacecraft positioned at 90° in the ecliptic plane and augmented by another one in a solar polar orbit should (see Figure 4.3) provide the necessary means to conduct these measurements. They should be equipped with coronagraphs, XUV and x-ray telescopes, particle detectors, magnetometers, and radiometers. The 1-AU spacecraft near Earth could be at L1; the Space Station could service an L1 platform essentially as well as a geosynchronous one.

Additional Solar and Heliospheric Studies

There is now a considerable body of evidence to suggest that all scales of structure on the Sun, as well as other astrophysically interesting objects, are ultimately governed by small-scale processes associated with intermittent magnetic fields and turbulent stresses. The understanding of the physics of the creation and decay of these dynamical structures is essential to a proper description of large-scale structures (such as coronal active regions, flares, and the solar wind) and their effects on interplanetary space and the near-Earth environment.

The interplay between processes occurring on vastly different spatial scales is ubiquitous in astrophysics. Whether in accretion disks feeding black holes at the center of active galaxies or quasars, in the magnetospheres of neutron stars, or in the x-ray coronae now known to surround a wide range of stars, small-scale magnetohydrodynamic processes are thought to influence and sometimes control the behavior of the object.

In these astrophysical situations, observations using even the most advanced technology currently conceivable will not allow us to directly observe the controlling small-scale processes. Using the Sun, however, we can indeed imagine direct observations. The Sun is therefore a unique tool for advancing our understanding of a broad class of astrophysical phenomena, if we can penetrate to the domain of underlying processes that often operate on spatial scales of 1 to 100 km.

An orderly progression of goals that could realize much of this promise would include the following:

1. Development of the successor to the Solar Optical Telescope and its integration into the Advanced Solar Observatory on the Space Station, along with the development of 0.1-arcsec ultraviolet and x-ray solar instruments on the Space Station;

2. Interferometric experiments in the ultraviolet and extreme ultraviolet, aimed at a preliminary reconnaissance of solar features at angular sizes much less than 0.1 arcsec.

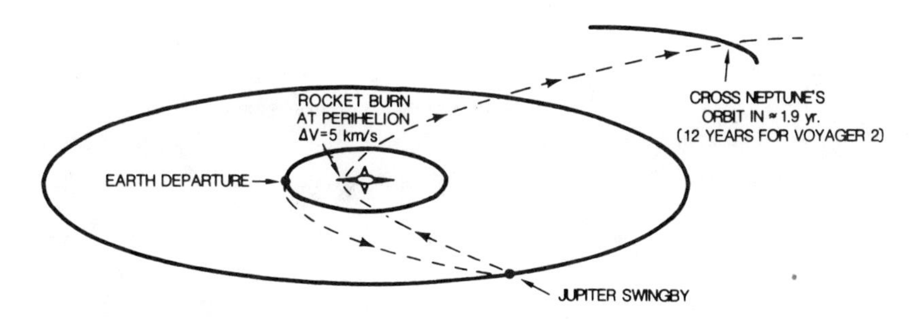

FIGURE 4.4 An interstellar probe using Jupiter gravity assist and solar swingby to escape the solar system. It would reach a distance of the orbit of Neptune in 1.9 years, in contrast to the 12 years required for Voyager 2 to reach the same distance.

3. Development of new 1-m class facilities, utilizing the emerging multilayer coating technologies, designed to obtain resolution in the 0.01-arcsec regime at extreme ultraviolet or soft x-ray wavelengths.

4. Improvement of the angular resolution of 1-m class telescopes by the use of multiaperture arrays to achieve baselines of order 10 m. Further details are contained in Appendix D.

Interstellar Probe

An Interstellar Probe, that could be launched about the year 2000, should reach beyond about 100 AU in a time interval of less than 10 years, preferably about 5 years. Several possible schemes, including Jupiter gravity assist and swingby of the Sun at 4 solar radii (see Figure 4.4) as well as use of megawatt nuclear electric propulsion, could provide the necessary acceleration for spacecraft velocities varying from about 50 to 100 km/s (11 to 21 AU/yr). The spacecraft should be instrumented redundantly with plasma, field, particle, and wave instruments, depending on detailed definition of science objectives and spacecraft capabilities. The fully instrumented spacecraft mass is likely to be in the range of 500 to 1000 kg.

The objectives of the mission are to determine the characteristics of the heliopause, interstellar medium, low-energy cosmic rays excluded from the heliosphere, and global interplanetary gas and mass distribution of the solar system, and possibly, a much more precise determination of the stellar and galactic distance scale through parallax measurements of the distance to nearby stars.

Some detailed plans for the science measurements were provided in a NASA report *An Interstellar Precursor Mission* (JPL 77-70, October 1977). A summary of the recommendations contained in that report follows.

Scientific Measurements

Heliopause and Interstellar Medium: Determination is needed of the characteristics of the solar wind just inside the heliopause, of the heliopause itself, of the accompanying shock (if one exists), and of the region between the heliopause and the shock. The location of the heliopause is not known; estimates now tend to center at about 100 AU from the Sun, but recent observations of low-frequency radio waves by the Voyager plasma wave investigation suggest even smaller distances.

Key measurements to be made include magnetic field, plasma properties (density, velocity, temperature, composition, plasma waves), and electric field. Similar measurements, extending to low energy levels, are needed in the interstellar medium, together with measurements of the properties of the neutral gas (density, temperature, composition of atomic and molecular species, velocity) and of the interstellar dust (particle concentration, particle mass distribution, composition, velocity). The radiation temperature should also be measured.

The magnetic, electric, and plasma measurements would require only conventional instrumentation, but high sensitivity would be needed. In situ measurements of neutral gas composition might require development of a mass spectrometer with greater sensitivity and signal-to-noise ratio than present instruments. Remote measurements of gas composition could be made by absorption spectroscopy, looking back toward the Sun. Of particular interest in the gas measurements are the ratios D/H, $H/H_2/H^+$, He/H, He^3/He^4; the contents of C, N, O, and if possible of Li, Be, B; and the flow velocity. Dust within some size range could be observed remotely by changes in the continuum intensity.

Cosmic Rays: Measurements should be made of low-energy cosmic rays, which the solar magnetic field excludes from the heliosphere. Properties to be measured include flux, spectrum, composition, and direction. Measurements should be made at energies below 10 MeV and perhaps down to 10 keV or lower. Conventional instrumentation should be satisfactory.

Pluto: If a Pluto flyby is contemplated, measurements should include optical observations of the planet to determine its diameter, surface and atmosphere features, and an optical search for and observations of any satellites or rings. Atmospheric density, temperature, and composition should be measured, along with charged particles and magnetic fields. Surface temperature and composition should also be observed. Suitable instruments include a TV camera, infrared radiometer, ultraviolet/visible spectrometer, particles and fields instruments, and an infrared spectrometer.

For atmospheric properties, ultraviolet observations during solar occultation (especially for H and He) and radio observations of earth occultation should be useful. The mass of Pluto should be measured: radio tracking should provide this.

TERRESTRIAL MAGNETOSPHERIC PHYSICS

Imaging of the Earth's Magnetosphere

The in situ and remote sensing observations of the ISTP Program are designed to provide information that can be used to construct a global model of the Earth's magnetosphere, and follow-on programs must be designed to test these models. The concept of imaging the terrestrial magnetosphere is currently being investigated, and the results seem very promising (see Appendix A). It appears that from a platform as far away as the Moon (L4, L5, or a lunar base) or L1 (250 earth radii), ultraviolet emissions from He^+ at 304 Å will be sufficiently intense from the high-density, low-temperature plasma regions of the plasmasphere, the magnetosphere, the geotail plasma sheet (during storms), and even the region of the bow shock (with long integration times) so that images could be constructed with reasonable resolution in time and space (approximately 100 km, approximately 10 min). Figure 4.5 shows a sketch of the magnetosphere, and it indicates the locations for the terrestrial imaging instruments. As noted on the figure, earth imaging and solar imaging can be carried out on low-altitude platforms (Space Station) as well as on these more distant bases. Within about 10 earth radii it is possible to "image" the more

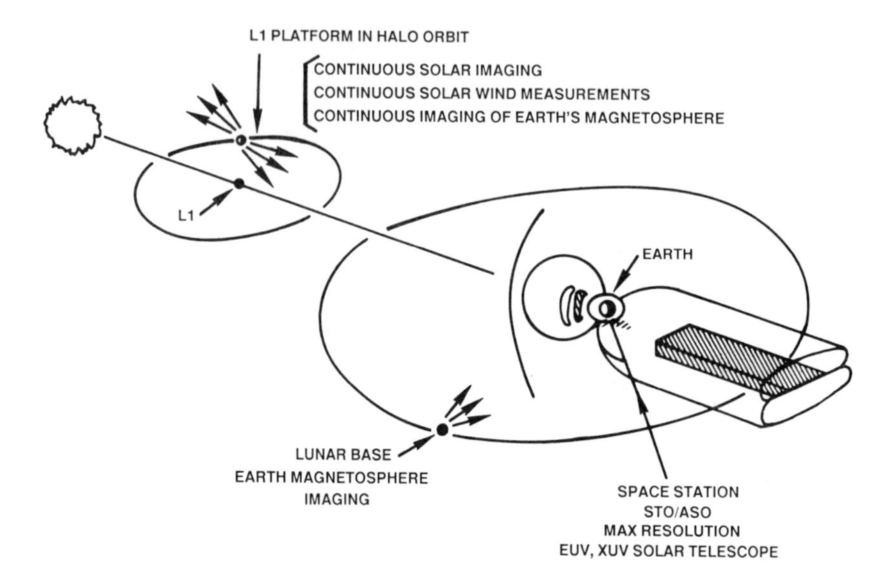

FIGURE 4.5 Locations of instruments for imaging the terrestrial magnetosphere: an L1 platform, a lunar base, and the Space Station.

energetic plasmas (e.g., ring current, plasma cusp) by observing energetic neutrals produced by charge exchange.

Plasma-Dust Interactions

Cosmic plasmas are often "dusty," by which it is meant that they contain solid particles, some of which are very small dust grains. These are usually electrically charged, and if their charge-to-mass ratio is large enough their motion may be dominated by electromagnetic forces so that they can be considered to be one component of a "dusty plasma." Charged dust in the solar system is known to be important for the origin of Jupiter's dust belt, for the fine structure of some Saturn's rings, and for the motion of both interplanetary and cometary dust. The population of dust grains in the Earth's environment appears to be rapidly increasing because of the injection of large amounts of exhaust in particulate form (mainly aluminum oxide) from solid fuel thrusters.

Interstellar gas clouds contain large quantities of dust, which plays an important role in determining the physical properties of the clouds. Photoelectron emission from grains by ultraviolet radiation can be a heating mechanism for the gas; conversely, dark clouds can be cooled by radiation from grains. In addition, the electron and ion densities in a cloud can be depleted by surface recombination on grains. Thus the gas pressure in a dusty cloud can be affected by the dust, with implications for the evolutionary collapse of a cloud and subsequent star formation.

The processes responsible for the production, growth, dispersion, and destruction of dust grains are poorly understood. It is difficult to study grain growth and the processes of coagulation and fragmentation and their dependence on grain charge in laboratories on Earth because of the dominance of gravity. Controlled experiments in space, either on dust in a spacecraft plasma chamber or on dust ejected from a spacecraft into a plasma environment, would greatly increase our understanding of these processes. A rendezvous mission of a spacecraft with a cometary dust cloud or planetary ring where the differential velocity of the spacecraft with the cloud was minimized would provide the opportunity for a detailed in situ investigation of both the dynamics and the physical processes that occur in such environments. It would be useful to develop spacecraft-carried remote sensing techniques such as radar or lidar scattering to obtain information on a macroscopic scale.

Several technology advances will be needed to study plasma-dust interactions. These include methods for manufacturing dust with the desired properties; methods for injecting dust clouds with the desired space, velocity, and charge distributions; methods for measuring the velocity, mass, and charge distributions of dust particles; and methods for the remote sensing of dust clouds.

Active Experiments: Gas-Plasma Interactions

Following the execution of gas releases in the solar wind, magnetosheath, and distant magnetotail by the Active Magnetospheric Particle Tracer Explorers (AMPTE) program and in the near-Earth magnetosphere from Combined Release and Radiation Effects Satellite (CRRES) in the middle and late 1980s, it is expected that studies of gas-plasma interactions will continue to expand in scope and complexity into the mid-1990s and beyond the year 2000. The objectives of this initiative are as follows:

1. To investigate the interaction between an artificially injected neutral gas and a cosmical plasma in the upstream solar wind, magnetosheath, plasma sheet, and auroral magnetosphere.

2. To trace the flow of plasma from the solar wind and the ionosphere through the entire magnetospheric system using plasma tracer techniques.

3. To modify the magnetosphere/plasma environment by means of plasma seeding in the equatorial magnetosphere and inducement of artificial auroras and magnetic substorms.

The objectives outlined above can be readily addressed by using gas-carrying spacecraft for injection/modification of the plasma environment in various areas of the near-Earth system. Spacecraft already developed for the AMPTE and CRRES programs and could be utilized with appropriate propulsion systems to make releases at different locations. Also, spacecraft carrying appropriate diagnostic instrumentation (Explorer class) should be distributed in various orbits throughout the magnetosphere for observing the tracer elements and charting their flow through the system. The choice of appropriate tracer elements may make possible remote sensing at very low levels of intensity, and thus provide an "image" of plasma motions in some regions.

The techniques for gas releases in space are well developed both in the United States and abroad. Similarly, sensor instrumentation necessary for determining the results of releases is currently in existence and reasonably adequate. Potential development of remote sensing instrumentation (visible, ultraviolet) for imaging the tracer plasma would substantially enhance the utility of these techniques in the future.

Injections of Plasma Waves and Particle Beams

All astrophysical plasma systems—stellar coronae and winds, planetary magnetospheres, interstellar media—support plasma waves and nonthermal particle beams. The frequencies of the plasma waves range from the Alfven (hydromagnetic) regime (which in many solar system plasmas can have periods as low as a few milliseconds), to ion and electron cyclotron waves (a few hertz to many kilohertz in planetary magnetospheres), to synchrotron radiation (a few tens of megahertz Jupiter's magnetosphere). Plasma waves are also produced in planetary magnetospheres by collective effects, including electrostatic plasma oscillations, and by lightning in planetary atmospheres.

Nonthermal particle populations can be produced by large- and small-scale electrostatic potentials and by energy conversion processes involving plasma instabilities. The propagation of such beams through an ambient plasma population can result in beam-plasma instabilities, which can radiate various plasma wave modes.

The physical interpretation of the plasma waves and nonthermal particle beams measured by spacecraft in situ in solar system plasmas requires thorough diagnostics of the existing plasma environment. Interpretations of the electromagnetic plasma radiations measured from remote astrophysical sources rely heavily upon the physical understanding achieved from detailed studies of solar system plasmas. An entirely new method of gaining understanding of naturally occurring plasma waves and particle beams is through the injection of man-made waves and beams into the natural plasma media.

Wave injection experiments into the magnetosphere have been conducted for some time from the ground. Recently, wave and beam injection experiments have begun using opportunities available with the Space Transportation System. Extended STS operations and, ultimately, the Space Station will provide the opportunities for more considered and extended injection experiments—taking into account the different ambient plasma conditions presented by different levels of geomagnetic activity. Higher power levels available from a Space Station for the injected beams and waves will permit study of highly nonlinear beam-plasma and wave-plasma systems, representative of a wide variety of natural systems.

An electrodynamic tether will be flown on a future STS mission. Again, a Space Station will provide the means for further and extended tether flights to study the injection of waves into an astrophysical plasma, waves with frequencies ranging from the Alfven regime to the very low frequencies regime. The results of such experiments will be of importance in their own right, from a basic plasma physics vantage point, and will also be of considerable relevance to astrophysical plasma systems containing large conducting, moving bodies, such as Io in Jupiter's magnetosphere.

TERRESTRIAL ATMOSPHERIC PHYSICS

Upper Atmosphere Science

The Earth's mesosphere and lower thermosphere are the least explored regions of the Earth's atmosphere. They are influenced by varying solar extreme ultraviolet (EUV), ultraviolet (UV), and x-ray radiation, auroral particles and fields, and upward propagating waves and tides from the lower atmosphere. There are strong interactions between the chemistry, dynamics, and radiation of both the neutral and the ionized constituents of this region. It is known that the global structure of this region of the atmosphere can be perturbed during stratospheric warmings and solar-terrestrial events (e.g., magnetospheric substorms, solar flares), but the overall structure and dynamic responses to these effects and even the basic controlling physical and chemical processes of these effects are not understood.

A comprehensive multisatellite mission is needed to measure the thermal, compositional, radiational, and dynamic structure of the mesosphere-ionosphere-thermosphere system. Remote probing of winds, temperature, and composition, combined with in situ measurements of thermospheric and ionospheric properties, will provide information on the dynamic processes, the depth of penetration into the atmosphere of solar effects, and the dissipation and transmission of waves and tides from the lower atmosphere. Two polar-orbiting spacecraft are needed to define interhemispheric differences in the response to solar-terrestrial events, and an elliptic orbiter is needed to probe the lower thermosphere and define interactions with the magnetosphere through auroral imaging and other measurements. In order to be able to make in situ measurement at the lowest possible altitudes (about 120 km), the tether facilities of the Space Station also need to be taken advantage of. An extensive network of ground-based radar and optical interferometers and spectrometers will determine time variations of atmospheric and ionospheric properties at given locations. Large numerical models of the general circulation, energetics, and chemistry of the thermosphere, mesosphere, and ionosphere will be used in the analysis and interpretation of data.

Space Station Atmospheric Studies

The Space Station and the co-orbiting and polar-orbiting platforms provide opportunities for extensive probing of the upper atmosphere and ionosphere. High-resolution interferometers, spectrometers, and radiometers can be used to determine detailed chemical, radiative, and dynamic properties of the middle and upper atmosphere and ionosphere, and are part of the proposed instrument complement of the Solar Terrestrial Observatory (STO) and the Earth Observation System (EOS). The tether provides an excellent opportunity to explore in situ the properties of the thermosphere and ionosphere from orbital altitude down to about 120 km, an altitude region that has been studied directly only intermittently by sounding rockets.

Global Current Missions

In a laboratory plasma the electric and magnetic fields are usually generated by circuits external to the plasma. A space plasma differs from a laboratory plasma in that the magnetic and electric fields and currents in the space plasma are self-consistently determined by the distribution of charged particles in the plasma. To understand the behavior of a space plasma, we must measure these fields and currents.

Current systems in magnetospheric plasmas may be distributed over large volumes, concentrated in sheets, or confined to flux tubes. The ring current in the Earth's radiation belt is a volume current. Examples of sheet currents are found at the magnetopause and in the center of the geomagnetic tail. The phenomenon known as a flux transfer event is principally a line current. Similar current systems are seen in planetary magnetospheres and ionospheres, the solar corona, and astrophysical systems, but the most accessible of these regions is the terrestrial magnetosphere.

Not only the distribution of current but also its temporal variation is important. Electric fields associated with the variation of electric currents and magnetic fields are responsible for the acceleration of particles to high energies. There is no way to measure these currents and fields remotely; in situ observation programs that probe the volume of interest are required. The existence of both large-scale and fine-scale current systems poses a programmatic challenge. A large-scale network is required with coarse resolution, and small clusters of probes with fine resolution that move in eccentric orbits through the lattice of the large-scale network are required. Possible configurations for these probes have been examined, and it is estimated that on the order of 300 probes are necessary for this investigation. These probes are identically configured to measure the magnetic and electric fields and electric current, except for a few that include selected charged particle measurements. This massive commonality of the systems used should enable the program to be undertaken for an acceptable cost.

PLANETARY SCIENCE

Mars's Aeronomy and Magnetosphere

At present we have explored the solar wind interaction with each of the planets out to Saturn, with the single exception of Mars. Thus, the magnitude of any intrinsic martian magnetic field remains uncertain, and the nature of the solar wind interaction is not known because the solar wind could be deflected by either a weak planetary magnetic field or the planetary ionosphere. Mass loading of the solar wind by the planetary ionosphere, the process responsible for the formation of cometary magnetotails, may or may not be important. We do not know whether the planetary atmosphere is shielded from the solar wind or whether it interacts strongly with the solar wind. Hence, we do not know the sources and losses of the martian ionospheric plasma and upper atmosphere.

While the Viking entry probes did carry retarding potential analyzers, those were insufficient to provide the global structure of the ionosphere and provided no data on the dynamics, energetics, or chemistry of the ionosphere and upper atmosphere. Thus, our understanding of the martian ionosphere lags far behind that of Venus. Because the state of magnetization of the martian ionosphere is expected to be different from that of Venus, we cannot simply extrapolate Venus data to Mars.

Currently approved missions do not address the problems of martian aeronomy and magnetospheric processes. The Mars Geoscience Climatology Observer is planned to carry a magnetometer. However, this one instrument is insufficient to address any of the outstanding problems of martian aeronomy, except the existence of an intrinsic field. Further, the low-altitude circular polar orbit is inappropriate for addressing the majority of the questions outlined above. The solutions require an elliptical orbiter carrying a complement of ionospheric and magnetospheric instruments such as neutral, thermal ion, and suprathermal ion mass spectrometers, thermal and suprathermal electron detectors, and magnetic and electric field and wave detectors. There are no technological developments required to undertake this mission. It is a prime candidate for an early flight in the planetary observer series.

Mercury's Magnetosphere

The dynamics of the terrestrial magnetosphere is strongly affected by field-aligned currents that close in the terrestrial ionosphere. These currents transmit stress from the outer magnetosphere to the ionosphere and thence to the neutral atmosphere. Mercury has no dynamically significant atmosphere or ionosphere and thus should respond very differently to the solar wind. In particular, ionosphere line-tying (which controls current systems in the Earth's magnetosphere) should be unimportant at Mercury and hence new classes of substorm mechanisms may develop there. We have few data on Mercury's magnetosphere. The existence of an intrinsic magnetic field was determined from two nighttime passes of Mariner 10. However, lack of knowledge of solar wind conditions during these passes limited the accuracy of the determination of the intrinsic dipole moment. The magnitudes of higher order moments have been determined even less accurately. The nature of the plasma circulation in Mercury's magnetosphere and its variation due to changes in solar wind conditions remains unstudied. Although energetic particle transients were found, the spectra were not adequately measured. Comparative studies of Mercury's magnetosphere are crucial to understanding the role of the terrestrial ionosphere in magnetospheric processes and to bridging the gap between the solar wind interaction with the Moon, which has weak, localized magnetization and has no atmosphere, and with the Earth, which has a strong magnetic field.

There are no currently approved missions to Mercury. A study of the intrinsic magnetic field of Mercury requires a low-periapsis polar orbiter and some knowledge of the strength of the solar wind. This could be provided by a single elliptical orbiter or multiple spacecraft.

The plasma, energetic particle, wave, and magnetic field measurements needed on this mission could be carried on a rather modest spacecraft. However, other discipline objectives could be accommodated on a large spacecraft; the trajectories permitting larger payloads should be determined. Further, studies of thermal input to the spacecraft as a function of orbital characteristics should be made.

One addition to the Mercury orbiter concept that may be especially valuable concerns relativity. The objectives are to determine the precession of perihelion of the planet with much improved accuracy; to determine whether G is constant with 10^{-13}/yr accuracy; to measure J_2 for the Sun directly; to improve measurement of the relativistic time delay and check for preferred frame effects; to improve knowledge of the gravity field of Mercury. These could be accomplished with a small transponder satellite in an orbit with a mean altitude of about 2500 km and eccentricity of less than 0.25. Two-frequency tracking from the Earth could determine the center-of-mass distance and provide improved information on the gravity field. The currently planned DSN Doppler tracking accuracy of 5×10^{-15} and improved ranging accuracy of 10 cm are required. Tracking data are needed for about 24 hours per week over a 2- to 8-year mission lifetime.

Jupiter's Ionosphere/Magnetosphere and Tail

The dynamics of the jovian magnetosphere and its ionospheric interactions are different from that of Earth because plasma from an internal source, the satellite Io, is known to dominate much of the behavior of the outer magnetosphere of Jupiter. The solar wind, on the other hand, is known to modulate intensities of some long-wavelength radio emissions. However, it is not known whether the jovian aurora is associated with the Io plasma torus alone or with the jovian cusp region, two logical possibilities that involve different energy transfer mechanisms. Practically no information is available on the structure of the jovian ionosphere and its interaction with the magnetosphere; the four occultation measurements of electron concentration, necessarily made at differing latitudes and jovian magnetic longitudes, are not mutually self-consistent. Furthermore, no direct information is available on the chemistry, energetics, and dynamics of the ionosphere and upper atmosphere. Meaningful advances in understanding the chemical and physical processes in a hydrogen/hydrocarbon-dominated upper atmosphere have to await such measurements. Finally, the magnetospheric tail, knowledge of whose dynamical behavior is important to understanding how the magnetosphere works, has not been explored at all on the dusk side, nor will it be explored during the Galileo prime mission. Without knowing how the tail behaves, particularly in the pre-midnight sector, we will not be able to understand how the magnetosphere operates as a system. Without knowing the structure of the ionosphere as a function of latitude and magnetic longitude, we will not be able to understand either the ionosphere or how it couples to the magnetosphere.

In order to address these questions, the task group suggests a Planetary Observer-class vehicle in an elliptical polar orbit about Jupiter, with a perijove of 500 km above the 1-mbar level, an apojove of 8 R_J from center of Jupiter, and a period of 28 hours.

For such an orbit, the major axis precesses approximately 1°/day and thus alternately is in a position either to view one of the poles or to pass through the Io torus. Measurement objectives include imaging of the aurora and the torus, determination of the plasma composition and density, and observations of energetic particles, plasma waves, radio emissions, and field configuration, as well as several ionosphere parameters. Further details are contained in Appendix B.

SUMMARY OF TECHNOLOGY DEVELOPMENT NEEDS

The technology development needs identified by this task group can be summarized as follows:

- Low-thrust propulsion (Solar Probe, Solar Polar Orbiter, Heliosynchronous Spacecraft, possibly Interstellar Probe)
- 4-solar radii heat shield (Solar Probe, Interstellar Probe)
- "Perihelion thruster" (Solar Probe, Interstellar Probe)
- Magnetospheric imaging techniques (L1 Platform, Space Station, Lunar Base)
- High-level radiation-resistant components (Jupiter Polar Orbiter)
- Lidar System for active probing of the atmosphere (Space Station)
- Ultra-low-cost current-measuring spacecraft (Terrestrial Magnetosphere)
- Active plasma physics experiments (interactions of plasmas with beams, waves, gases, and dust) (Space Station)
- High-reflectivity multilayer coating for EUV and XUV (Space Station)
- Enhanced dust impact protection (Jupiter Polar Orbiter)

5
Summary of Technology Development Needs

SOLAR AND HELIOSPHERIC

1. Solar and heliospheric physics requires in situ plasma measurements from close to the solar surface to the interstellar medium. Conventional propulsion cannot be used for the following orbits: (a) an elliptical orbit for the solar probe with a 1-year period; (b) a circular or near-circular orbit for the heliosynchronous satellite at 30 solar radii; and (c) at least a 40-km/s velocity for a spacecraft leaving the heliosphere.

Solar Electric Propulsion (SEP) or similar methods need development to have a technology available by 1995. It should be noted here that low-thrust propulsion is also needed for many comet and asteroid rendezvous missions.

2. In connection with spacecraft coming close to the Sun, thermal isolation techniques need to be investigated. For the interstellar probe, one must develop nuclear electric propulsion or a perihelion thruster.

3. Multilayer coating will allow the construction of normal incidence mirrors for the x-ray/ultraviolet and x-ray regime that will result in unprecedented resolution. Present-day techniques are not applicable to large-diameter (about 1 m) mirrors. Active

mirror surface control techniques and new telescope stabilization methods need to be researched.

4. A Lagrangian platform would be of use to many disciplines. It should be studied in terms of instrument volume, weight, power, and serviceability.

MAGNETOSPHERIC PHYSICS

1. Systems will need to be developed that will allow the plasma populations to be "imaged" so that global models of magnetospheric structure can be tested directly.

2. The development of radiation-resistant sensors and electronic components that can extend the lifetime of the Jupiter Polar Orbiter mission is needed.

3. In order to be able to deploy a large network of spacecraft to map current systems in the magnetosphere, the costs must be tightly controlled. Cost-reduction techniques should be investigated for the development of simple identical spacecraft.

4. Since the Jupiter Polar Orbiter spacecraft will necessarily fly through the jovian ring system, enhanced dust protection techniques will have to be developed.

5. Techniques and systems needed to carry out active plasma physics experiments (interactions of plasmas with waves, beams, gases, and dust) on Shuttle/Spacelab and/or Space Station should be developed.

Appendixes

The appendixes that follow are reports or excerpts of reports that resulted from workshops conducted by NASA in support of the present study. These efforts are an important part of the study and are included here in order to make the task group report complete.

The four workshops were conducted by different organizations and in different manners. Thus, the resulting reports are very different in character. What follows are four documents that represent, at least, the essential features of these reports:

A. Workshop on Imaging of the Earth's Magnetosphere. This workshop was convened by the task group at the NAS Woods Hole study center during the task group meeting of July 1985. Appendix A represents the complete report.

B. Jupiter Polar Orbiter Workshop. Appendix B is a summary of the report of the workshop held at UCLA in July 1985.

C. Workshop on Plasma Physics Research on the Space Station. Appendix C is taken from a review paper describing the results of the workshop held in May 1985 in Alabama.

D. High-Resolution Observations of the Sun. Appendix D represents the complete report of the workshop held at the National Solar Observatory in Tucson, Arizona, in January 1986.

Appendix A
Workshop on Imaging of the Earth's Magnetosphere

WORKSHOP PARTICIPANTS

Lyle Broadfoot, University of Arizona
Andrew Cheng, Johns Hopkins University/Applied Physics Laboratory
Paul Feldman, Johns Hopkins University
David Gorney, Aerospace Corporation
Warren Moos, Johns Hopkins University
Edward Roelof, Johns Hopkins University
Donald Shemansky, University of Arizona
Donald Williams, Johns Hopkins University/Applied Physics Laboratory

CONTENTS

A.1 INTRODUCTION

Our present knowledge about the structure of the Earth's magnetosphere has been formed over the years largely on the basis of local measurements from spacecraft that move along trajectories that are isolated in space and time. However, even these limited observations show that the magnetosphere is essentially a dynamic system with large configuration changes driven both by direct variations in the solar wind and by more indirect processes that trigger large-scale instabilities associated with substorms.

Because of their enormous sizes and because there are large-scale variations in both space and time, planetary magnetospheres pose major challenges to scientists attempting to understand global behavior—an understanding that is necessary in a program aimed at developing quantitative predictive models of these systems.

Since single-point observations from an isolated satellite within the magnetosphere are obviously inadequate for this task, the initial response to the challenge of studying global behavior has been to conduct simultaneous multisatellite observations within the magnetospheric system. For example, the International Magnetospheric Study (IMS) program was based on use of the ESA GEOS satellites in geostationary orbits, the NASA ISEE-3 spacecraft at the sunward libration point, and the NASA-ESA spacecraft ISEE-1 and ISEE-2 in the same elliptical orbit with apogee of 23 earth radii. This program was highly successful, and it provided definitive information on the structure of various thin boundaries and on localized plasma physics phenomena—a success due to the small separation between ISEE-1 and ISEE-2 and due to their coordinated observing schedules. However, the program was not nearly as successful in its thrust toward studies of global dynamics. This was due both to the lack of coordinated observing schedules for the primary IMS spacecraft and to the fact that the satellite trajectories were not chosen to optimize global studies.

Thus, even after the IMS, a number of different theories existed to explain global observations, but definitive choices could not always be made on the basis of available data. One example is the controversy between boundary layer and reconnection models of substorm-related phenomena in the Earth's magnetotail. Another example is the question of whether energy is stored for long periods in the magnetotail and then released suddenly to cause a substorm, or whether a substorm is caused by a transient increase in energy input from the solar wind.

Individual components of the magnetosphere such as the field-aligned current systems near the Earth, the plasma sheet distributions, and the high-energy radiation belts have been relatively well measured separately but are poorly understood as interacting parts of a whole. Fundamental controversies on the nature of global dynamic phenomena such as substorms are a natural result. Indeed, significant advances have recently occurred in one of the areas cited above—aurora and field-aligned currents near the Earth—precisely because of the advent of global auroral imaging in the ultraviolet with simultaneous charged particle and magnetic field measurements on the Dynamic Explorer mission. Examples of these very significant observations are shown in Figure A.1.

The International Solar-Terrestrial Physics (ISTP) program represents a major new thrust in our attempt to understand global magnetospheric dynamics. This program, specifically designed for global studies, will place appropriately instrumented satellites in key magnetospheric locations, and simultaneous auroral imaging will be utilized to obtain data on the magnetospheric energy deposition into the atmosphere. A strong theory, modeling, and computer simulation effort has been incorporated into ISTP from the beginning, thus assuring the development of global models to be used with the ISTP data base. Finally, since ISTP has been designed as a global studies program from the outset, the participating space agencies (NASA, the European Space Agency, and the Japanese Institute for Space and Astronautical Sciences) have agreed on full coordination of observing schedules to maximize simultaneous coverage from the ISTP spacecraft. They will also establish a common central data base facility through which all ISTP researchers will have access to all ISTP data.

A major result of the ISTP program will involve the development and initial qualitative testing of the first truly global model of the Earth's magnetosphere. More accurate tests will certainly be needed as the community attempts to move the modeling effort into a quantitative predictive phase. Ordinarily this requires a dense grid of observing platforms. However, for a system as vast as the terrestrial magnetosphere, it would not be economically feasible to provide a sufficiently dense grid using local spacecraft observations alone. A new and innovative approach is required,

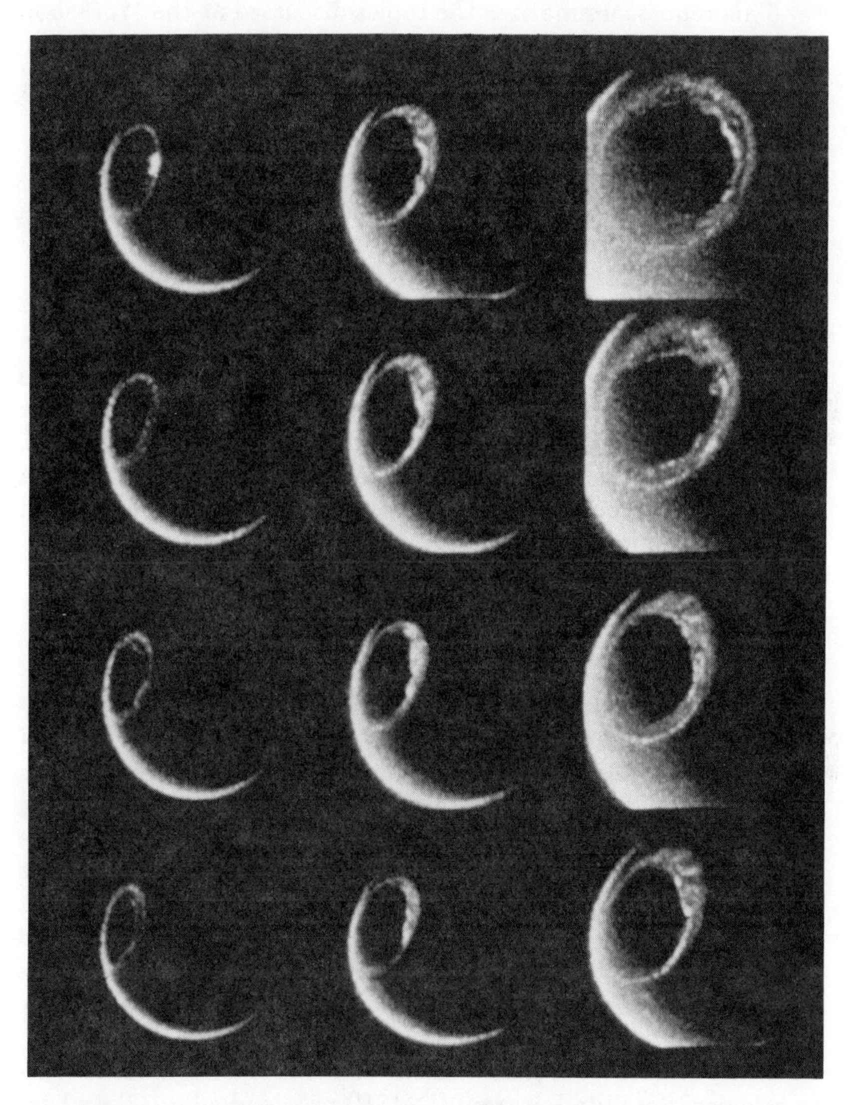

FIGURE A.1 Ultraviolet auroral images from Dynamics Explorer.

and recent measurements suggest that this approach should involve development of an imaging capability for the magnetosphere as a whole.

This report summarizes the topics discussed at the Workshop on Imaging of the Earth's Magnetosphere, which was held at the NAS Study Center in Woods Hole, Massachusetts, on July 30 and 31, 1985, under the auspices of the Solar and Space Physics Task Group of the Space Science Board's study: *Space Science in the Twenty-First Century.*

In the next section is a brief description of the science return from the present programs involving topside imaging of the auroral oval at atmospheric altitudes and expectations for improvement. The third section contains a discussion of the prospects for extreme ultraviolet imaging of the relatively dense and cool plasma of the plasmsphere, magnetopause, bow shock, and storm-time plasma sheet. The use of energetic neutrals (generated by charge exchange) for imaging of high-temperature plasma regions (e.g., the ring current and outer radiation belt) is discussed in the fourth section, and the conclusions and recommendations are contained in the final section.

A.2 IMAGING OF THE AURORAL OVAL: PRESENT STATUS AND FUTURE NEEDS

Auroral zone observations of visible, ultraviolet, and x-ray emissions generated by the precipitation into the atmosphere of energetic particles streaming downward along the geomagnetic field have provided extremely important means for remote sensing of dynamical phenomena that develop in the magnetosphere. The auroral region has been compared to a television screen that allows us to view the end effects of very remote magnetospheric processes. For years we looked upward toward the aurora from the ground and we were able to observe parts of the auroral displays, but the ground-based observations were limited in space and they were also restricted to the visible part of the spectrum.

Early results from the low-altitude polar-orbiting Defense Meteorological Satellite Program (DMSP) provided a large data base of broadband visible auroral imagery. From this data set much was learned about the large- and small-scale morphology of the aurora, the response to changing conditions in the solar wind, and the occurrence of brief but energetic substorms. These early photographs from space were also very significant because they provided a medium to tie together other satellite and ground-based observations of the particles and fields responsible for and resulting from the aurora. A major drawback of these early observations was their scarcity—the DMSP orbits provide auroral images with temporal separations of 90 min.

The recent Dynamics Explorer (DE) satellite remedied this problem by imaging the aurora continuously from a high-altitude elliptical orbit. Figure A.1 shows a set of examples from DE, taken consecutively at ultraviolet wavelengths. Auroral images of this type are also available with somewhat diminished quality under sunlight illumination (as demonstrated by observations from the low-altitude HILAT spacecraft). We now have continuous global images from the auroral region, and they have yielded extremely significant new information on auroral forms and morphology.

Scientifically, auroral imagery could provide even greater quantitative information if simultaneous multispectral images were available. Such multispectral techniques are now being developed to derive quantitatively the characteristics of the perturbations of the neutral and ionized atmosphere due to the aurora.

X-ray imaging of the aurora from space also offers promise for quantitative remote sensing of low-altitude processes since the x-ray emissions (bremsstrahlung) occur directly as a result of energetic auroral electrons colliding with atmospheric constituents. A number of successful imaging x-ray instruments have been flown in low-altitude polar orbit. Although the low-altitude observations have provided an opportunity to test the analysis techniques and instrumentation, they have not provided an adequate combination of spatial resolution, energy resolution, and aperture to perform true quantitative global imaging of the aurora.

The full potential of x-ray imaging demands instrumentation based at high altitude in order to view the entire polar region. The system must also be capable of providing adequate measurements of x-ray spectra on reasonable time scales and Figure A.2 shows the integration time needed to produce a quantitative x-ray image as a function of altitude and spatial resolution within the aurora; here an aperture of 100 cm^2 is assumed.

The ISTP polar spacecraft will carry multispectral auroral imagers that will cover visible, ultraviolet, and x-ray wavelengths, and the payload will also include advanced instrumentation for measurements of local plasma physics phenomena.

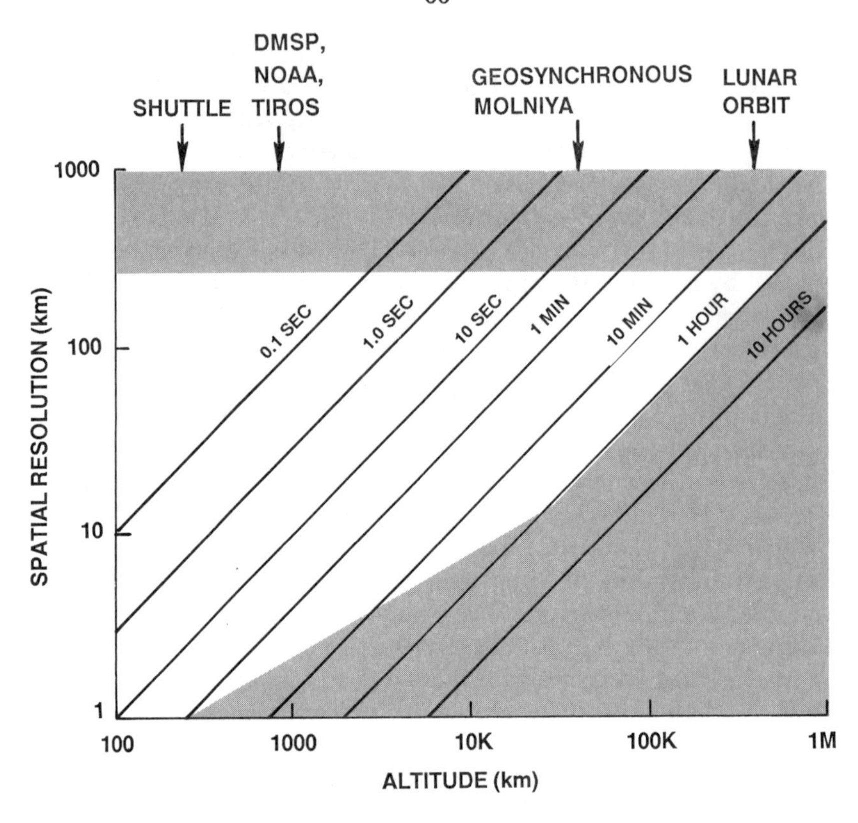

FIGURE A.2 Temporal and spatial resolution as a function of distance from source to x-ray detector. Required integration time A = 100 cm^2.

A.3 PROSPECTS FOR EXTREME ULTRAVIOLET IMAGING OF PLASMA IN THE MAGNETOSPHERE

Plasma temperatures above about 40,000°K generally produce ions with resonance transitions into the EUV region of the spectrum. In this temperature regime, energy lost by radiative processes is then dominantly in the EUV, and study of emissions at the short wavelengths can yield accurate diagnostic information about the plasma. An example of EUV plasma imaging involves the data provided by the Voyager spectrograph as it detected the plasma torus of Jupiter's moon, Io. Successive sweeping of the field of view of this one-dimensional imaging system has yielded the three-dimensional image of the jovian system shown in Figure

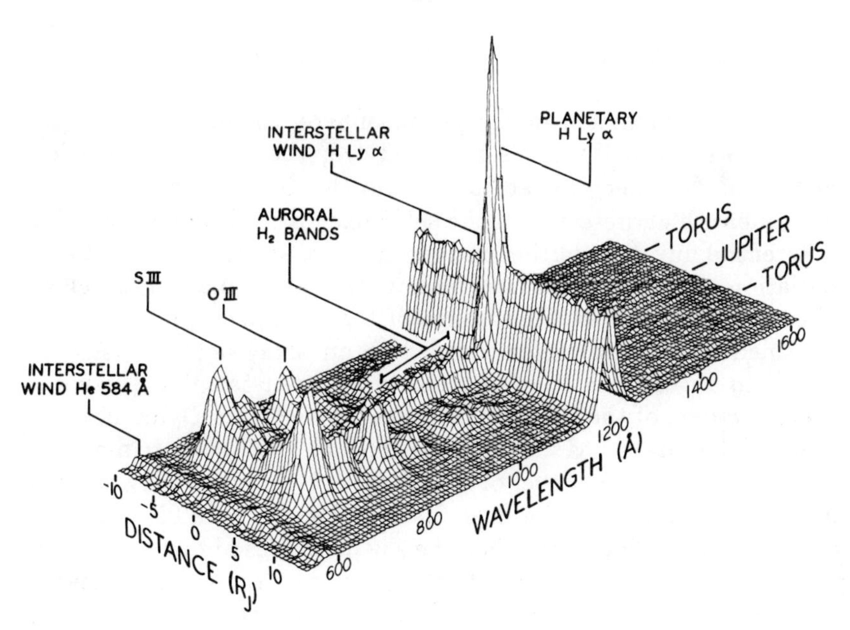

FIGURE A.3 Spatially and spectrally resolved image of Jupiter and the Io plasma torus obtained by the Voyager EUV spectrometer.

A.3. These observations have provided many pictorial views of the integrated emission from the plasma torus. The measurement of the characteristics of the EUV source (size, shape, spectral intensity, etc.) allows the identification of the ion species, and the determination of the number density and electron temperature distributions.

On the basis of these points of reference we can make a reasonable determination of our ability to image the terrestrial magnetosphere with EUV radiation. It is currently technically feasible to measure emission brightness at the level of 0.01 Rayleigh, within a reasonable time frame of about 100 s. On this basis an image of Earth's magnetosphere in the resonance fluorescence of the He^+ 304 Å line, for example, would provide an isophote map to a distance of 10 R_e geocentric. Arguments at a similar level can be made for the detection of other species in emissions produced by collisions and by fluorescence. The workshop participants conclude that the mapping of the plasmasphere and magnetosphere in EUV emission is an obtainable goal.

Plasmaspheric and Magnetospheric Boundaries

The dominant ion in the plasmasphere capable of producing EUV emission is He^+. As noted above, the principal emission feature of this species is at 304 Å, and this is produced primarily by resonance fluorescence of the solar line. Knowledge of the solar differential flux distribution and the spectral shape of the scattered radiation allows diagnostics of bulk motion of the plasma as well as plasma temperatures. The image of the terrestrial plasmasphere obtained by a photometer coupled to an array detector is illustrated in Figure A.4; these profiles have been calculated using a specific model of the He^+ distribution derived from Dynamics Explorer observations, as shown in Figure A.5. The ability to measure EUV emission at the level of 0.01 Rayleigh in an imaging system is entirely possible in the near future. Recent AMPTE-CCE observations of He^+ suggest that the magnetopause boundary could also be imaged. Moreover, it is known that the distant plasma sheet is populated by He^+ (and O^+) during storms, and hence imaging of dynamical tail phenomena during substorms should also be possible.

Other less extensive species having lower abundances, such as O, O^+, N, and N^+, could be observed with an imaging spectrometer and would provide measurement of the nature of the interface between the magnetosphere and the ionosphere. In general, these species are excited by both plasma electrons and fluorescence of solar lines. Spectral analysis of the emissions from these species are diagnostic of basic plasma properties; composition, density, and temperature of both ions and electrons.

The analysis technique has been used in astrophysics for many years (for instance, in the study of gaseous nebula), and it is equally suited to the study of the plasma environments of Venus, the outer planets, and comets, provided that suitable atom and molecular transitions are chosen for each particular system. A column abundance of $10^{10}/cm^2$ of OII ions interacting with electrons having typical auroral energies could be detected by present-day instrument design with a 100-s integration time, for example. Given somewhat larger integration times, processes at the bow shock and magnetosheath should be observable. Bulk motions in the magnetotail may be traced through successive measurements of structural features and through high-resolution spectral line shape measurements.

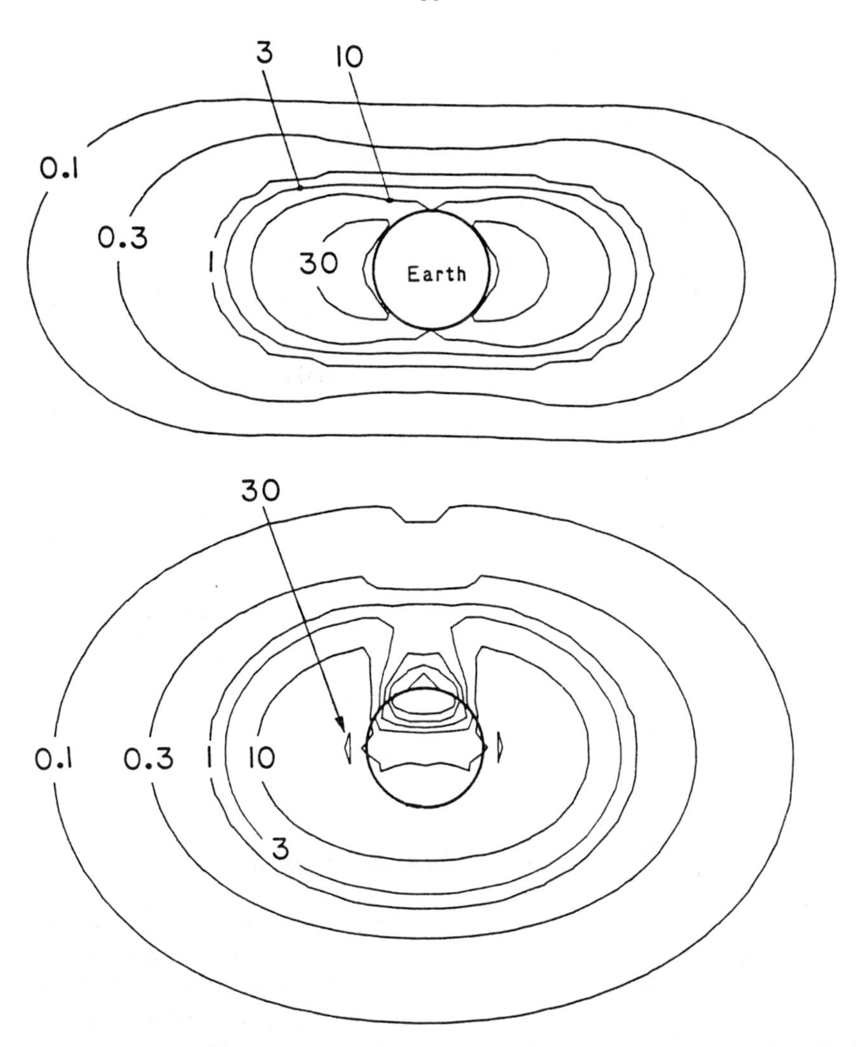

FIGURE A.4 He^+ 304 Å isophotes for distribution model 1 (see Figure A.5), as seen from two views: (top) a point on the Earth-Sun line in the plane of the magnetic equator, and (bottom) a point on the noon meridian at a magnetic latitude of 35°.

These global observations of the magnetosphere must clearly be linked to direct parallel measurements of solar activity, and followed through a time scale of one or more major solar cycles.

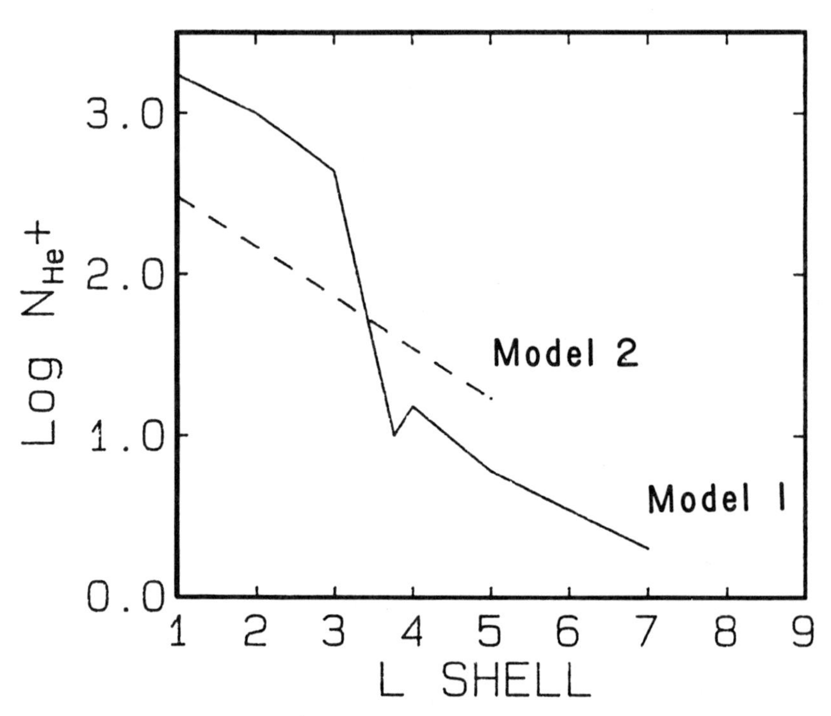

FIGURE A.5 Model He^{+} distributions derived from Dynamics Explorer observations.

Locations for Observing Stations

Earth's magnetosphere can be imaged from a range that allows the whole system to be continuously in the field of view of the imager. This can be done from the L1 Lagrangian point, as a control monitor, with spacecraft at the L4 and L5 Lagrangian points or an instrument on a lunar base to provide stereoscopic capability.

The Moon and the L4 and L5 points have many good features as sites for these measurements. The L4 and L5 points are at lunar distance, preceding and following the Moon at 60° to the Earth-Moon line. The 120° aspect angle is almost ideal for stereoscopic studies of discrete plasma motions of detached plasmas. The lunar orbital rate is slow enough to allow correlated studies over several days. The time scale is also about right for the study of typical solar-magnetospheric interactions. The changing aspect of the lunar orbit is important for viewing interface regions such as the magnetopause and bow shock surfaces. These interfaces need to be investigated with long optical paths that occur when the viewing direction is nearly tangent to the surface. For instance, the bow shock should be viewed tangentially over an arc-shaped region of the sky. The tangent line progresses so as to result in a complete survey of the complete bow shock in each lunar period.

Simultaneous imaging of the magnetospheric system from the L1 point is also required. This is an important viewing station because whereas L4, L5, and the Moon orbit the Earth, the L1 view is stable and stationary with respect to the Sun's direction. The solar wind can also be measured continuously from the L1 station, allowing detailed study of the effect of solar wind variability on the magnetospheric system. From here one could also study changes in the solar flux spectrum and intensity; this information is of importance for atmospheric analysis, and it is also needed for the complete interpretation of the resonance emissions from the magnetospheric plasma.

Lidar

Lasers can be used to excite emissions from the constituents of the ionospheric and magnetospheric plasma, as well as the atmosphere. This lidar technique involves the excitation of ions and atoms by a light source with highly controlled characteristics, and this, in turn, leads to precise information on species concentration, location, and velocities. NASA is developing such a lidar system for remote sensing of the middle atmosphere from the Space Shuttle and Space Station, with planned operations in the 1990s. Although the densities of magnetospheric particles are very low in comparison with atmospheric densities, the path lengths are long, and therefore acceptable signals may be generated. The major obstacle to applying this technique to magnetospheric remote sensing is the lack of a laser with reasonable efficiencies to operate at ultraviolet wavelengths. However, DOD requirements could lead to development of an ultraviolet laser, and NASA should be prepared to take advantage of such advances and utilize laser technology for remote mapping of the magnetospheric plasma.

A.4 PROSPECTS FOR NEUTRAL PARTICLE IMAGING OF PLANETARY MAGNETOSPHERES

Background

A new window into magnetospheric physics has been opened with the detection of energetic neutral particles (50 keV) emanating from the magnetospheres of Earth, Jupiter, and Saturn. These measurements directly point the way toward an innovative class of instruments devoted to global imaging of magnetospheric neutral particle emissions. Energetic neutrals are created within the magnetosphere by charge exchange reactions between fast magnetospheric ions and ambient neutral atoms or molecules (see Figure A.6). Since the resulting fast neutrals escape from the magnetosphere on rectilinear trajectories, they can be used to image the neutral-particle-emitting regions of the magnetosphere. The resulting images provide the only known means for remote sensing on a global scale of the magnetospheric energetic charged particle populations.

In this way energetic neutral images are complementary to ultraviolet and optical images of the magnetosphere, which remotely sense only the low-energy (~5 keV) charged particle populations. Unlike conventional spacecraft observations of charged particles that are essentially single-point measurements that sample only very small regions at any time, energetic neutral particles can image the entire complex magnetospheric system at a single instant. Charged particle measurements sample only a region comparable to the mean free path if scattering occurs; otherwise they sample only a flux tube of radius comparable to the gyroradius. Both length scales are much smaller than global length scales. Some very low frequency plasma waves are observed after propagation from global distances, but these are of such long wavelength that little or no directional information can be obtained.

Expected Results

Global imaging of neutral particle emissions from Earth's magnetosphere will yield a completely new global view of dynamic processes such as changes in plasma sheet configuration, growth and decay of the ring current, and auroral zone charged particle precipitation. Imaging studies of the interrelationships among

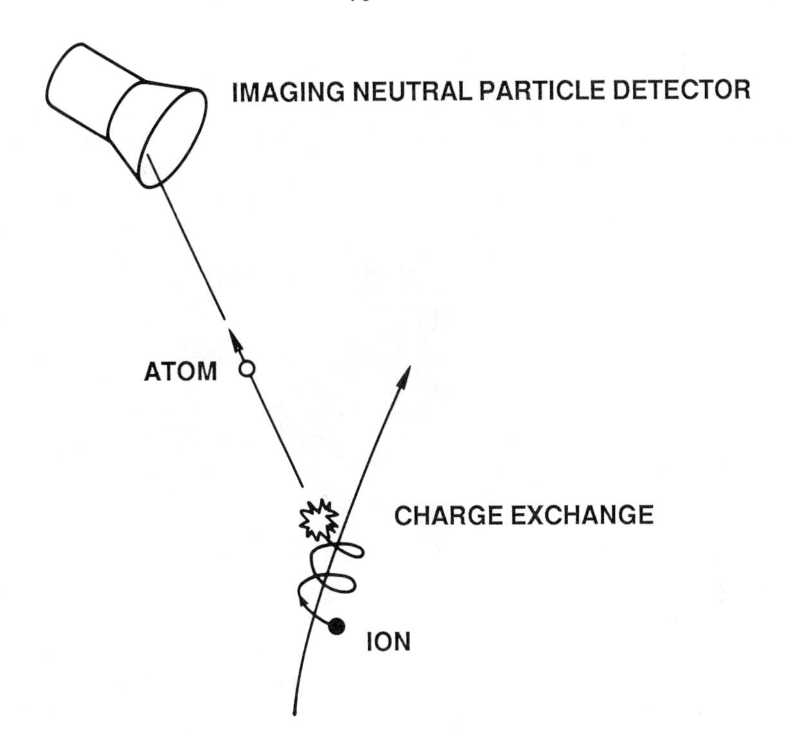

FIGURE A.6 Energetic neutral particle imaging of the fast ion and ambient neutral population. An energetic trapped ion captures an electron from an ambient neutral in a charge exchange reaction, becoming an energetic neutral atom, which then escapes the magnetosphere to the detector along a direction determined by the ion's velocity at the time of the reaction.

these processes on a global scale should resolve the long-standing debates concerning the nature of geomagnetic substorms.

Coarse global images of Earth's ring current using neutral particle emissions have already been obtained. Analysis of ISEE-1 and IMP 7,8 data has revealed energetic neutral particle emissions from charge exchanges between ring current ions and hydrogen atoms of Earth's geocorona.

Figure A.7 is an example (constructed from an eight-pixel ISEE-1 image near the midnight meridian at 20 R_e) of the energetic neutral particle emission during the recovery phase of a magnetic storm; the emission comes from the magnetic equatorial region of the ring current ($3 \leq L \leq 5$), and it is strongly asymmetrical, being concentrated in the dusk-midnight section.

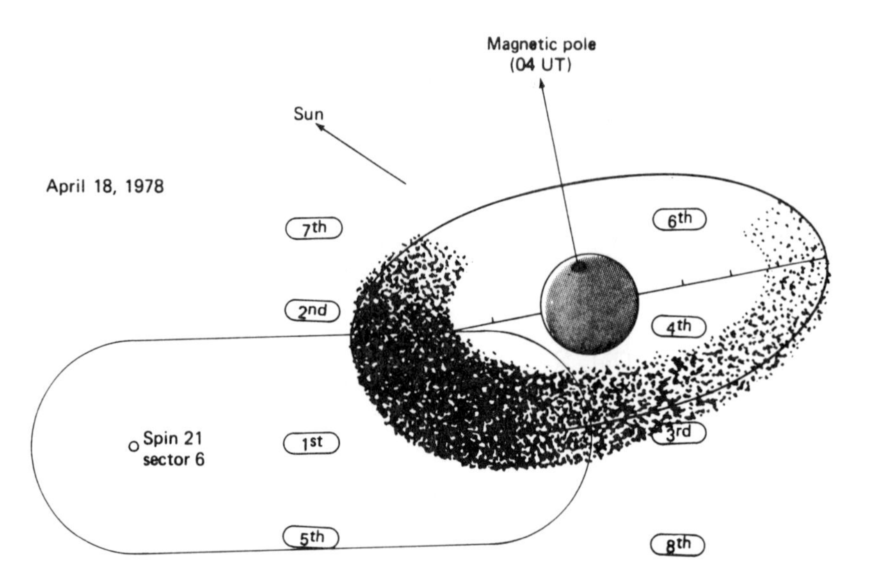

FIGURE A.7 Energetic neutral particle emission pattern during the recovery of a geomagnetic storm ISEE-1 near the midnight meridian at 20 R_e. Ranking of the pixel intensity (in lozenges marking the center of each pixel) indicates the greatest intensity from the late evening quadrant of the magnetic equatorial region ($3 < L < 5$).

Figure A.8 shows how a high-resolution neutral imager could provide much more information. The top panel contains the "image" expected from an ideal energetic neutral atom detector located at $X = -8R_e$, $Y = 0$, $Z = 5R_e$. For simplicity a dipole magnetic field, a nearly isotropic equatorial pitch angle distribution with empty loss cone, and an azimuthally symmetric ring current are assumed. The bottom panel of Figure A.8 then shows the line-of-sight ion flux column density for the main phase ring current distribution deduced from the measured neutral atom intensity profile in the bottom panel.

Energetic neutral particles also contain specific information on the composition of the magnetospheric energetic ions. For example, from Table A.1 the decay time and the energy spectrum of the neutral particle emissions observed during the recovery phase of a small magnetic storm are given; here it was deduced that O^+ was a significant component of the ring current. This illustrates the strong need for composition resolution capability in this new class of instruments.

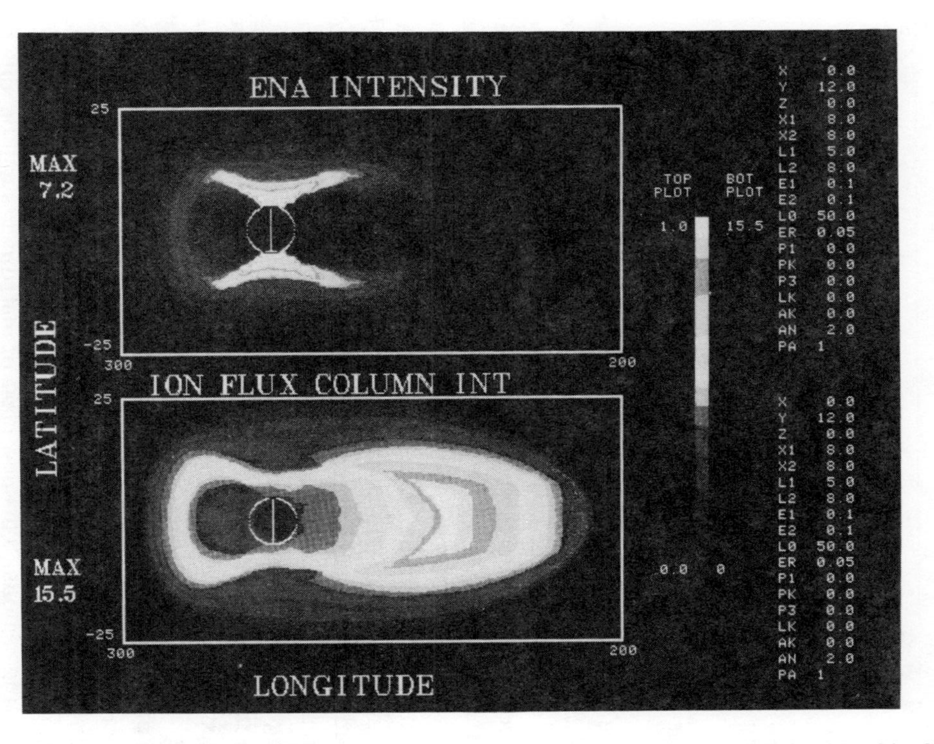

FIGURE A.8 (top) Expected energetic neutral atom intensities associated with this distribution. The energetic neutral measurements allow one to reconstruct the ion distribution. (bottom) Ion flux density for a main phase ring current distribution.

TABLE A.1 Nominal Neutral Particle Intensities from Various Components of Earth's Magnetosphere

	E ≳ 40 keV Proton Intensity (cm^2 s sr keV)$^{-1}$	Hydrogen Column Density (cm^{-2})	Energetic Neutral Intensity (cm^2 s sr keV)$^{-1}$	Count Rate per Pixel, Angular Resolution
Near-Earth Plasma Sheet	10^4	10^{10}	0.01	0.1/sec, -10°
Quiet Time Radiation Belts	2×10^4	5×10^{11}	1[a]	0.5/s,[a] -2°
Storm Time Ring Current	3×10^5	5×10^{11}	15[a]	7.5/s,[a] -2°

[a]Energetic neutral intensities and count rates up to an order of magnitude greater are predicted if O^+ dominates rather than protons.

The neutral particle intensity from charge exchanges in the Earth's magnetosphere is estimated from the integral along the line of sight of the product of the ion intensity, the charge exchange cross section, and the neutral density. Table A.1 gives nominal estimates of neutral particle intensities at $E \sim 40$ keV from the near-Earth plasma sheet, the quiet time radiation belts, and the storm time ring current. The first of these refers to an equatorial line of sight near 10 R_e through the plasma sheet. Both quiet time radiation belt and storm time ring current refer to lines of sight near $L = 4$. A 20-keV energy band is assumed, and a charge exchange cross section equal to 10^{-16} cm^{12} (appropriate for 50-keV protons incident on hydrogen) is used. The charge exchange cross section and predicted neutral intensities are about 10 times greater if O^+ ions are assumed to dominate protons, since the charge exchange cross section is 10^{-15} cm^2 for ~100 keV O^+ ions near incident on hydrogen.

Nominal count rates per pixel are shown in Table A.1. These rates apply to pixels whose fields of view are filled by the respective emitting regions. This new class of instruments will be able to image the quiet time radiation belts and the storm time ring current with high angular resolution (2°) and high time resolution (13 to 200 s). They also image the near-Earth plasma sheet at lower resolution. These studies will revolutionize our understanding of global dynamic processes such as magnetic substorms, because we will, for the first time, be able to see the entire magnetosphere in one image. Energetic neutral particle imaging can revolutionize our understanding of planetary magnetospheres in general. The Voyager spacecraft detected energetic charge exchange neutrals from both Jupiter and Saturn, but the instruments could not produce images. At Jupiter, energetic neutral imaging fulfills a unique role. It is the only known way of continuously sampling the very intense charged particle population in the innermost region of the jovian radiation belts, where particle detectors are saturated (and heavily damaged) by the penetrating radiation. We still do not know if there are magnetic storms at Jupiter and Saturn (or Uranus) similar to those at Earth. Global images of their energetic ion populations offer the means to answer this question—and perhaps to discover a new generalization of the concept of magnetic storms. Remote sensing, particularly from a high-inclination orbit, could be compared directly with earth-based optical and ultraviolet torus images and radio observations of synchrotron radiation.

A.5 CONCLUSIONS AND RECOMMENDATIONS

The exploration of the Earth's environment has been highly incomplete in the sense that we have not yet developed the technology to determine the complete dependence of atmospheric and magnetospheric dynamics on solar activity. Our understanding of the behavior of the Sun's emissions over a solar cycle is rudimentary, and our knowledge of the global response of the Earth's magnetosphere on both a short- and a long-term basis is very incomplete.

The workshop participants have investigated the future needs of the discipline and conclude that the full ISTP program will provide an excellent data base for the development of one or more dynamic global models of the Earth's magnetosphere. They also conclude that in the post-ISTP era, a new approach will be needed to ensure that the correct model is adopted, and to verify its accuracy.

The workshop discussions strongly suggest that the only practical post-ISTP approach to testing of global magnetospheric models is one that utilizes techniques to provide global images of the magnetosphere. Fortunately, the workshop deliberations also suggest that this imaging concept is a realistic one. It was demonstrated (using Voyager data from Jupiter) that atomic lines are excited by ambient electrons with sufficient intensity to provide images of cool dense plasma boundaries. For Earth the He^+ 304 Å line appears to be the best candidate for imaging the plasmasphere, the bow shock and magnetopause region, and (during storm conditions) the plasma sheet in the geomagnetic tail; however, this is currently a theoretical concept, since a terrestrial He^+ 304 Å imager has not been flown.

The more energetic plasmas (ring current, auroral region, and so on) are best "imaged" by searching for energetic neutral atoms produced by charge exchange. The very low resolution results already obtained with the ISEE-1 energetic particle analyzer verify the concept and show the power of this technique, but they also point to the need for development of instruments with higher resolution.

The workshop participants urge NASA to support the development and testing of suitable sensitive high-resolution magnetospheric imaging instruments with the aim of establishing a full-fledged magnetospheric imaging mission in the post-ISTP time frame.

Appendix B
Excerpts from the Final Report of the Jupiter Polar Orbiter Workshop

B.1 MOTIVATION, GOALS, AND CONCLUSIONS AND RECOMMENDATIONS

The inner jovian system comprises the inner Galilean satellites, Io's heavy-ion torus, the radiation belts, Amalthea, Jupiter's rings, and Jupiter itself. This region contains the solar system's most intriguing collection of phenomena and processes relating to planetary and satellite interiors, surfaces, atmospheres, ionospheres, magnetospheres, and rings.

Believing that no spacecraft could survive the radiation environment there long enough to make sufficient measurements, NASA has scheduled no mission to the inner jovian system. Recently the Committee on Planetary and Lunar Exploration, chaired by Donald Hunten, and the Space Science in the Twenty-First Century Task Group on Solar and Space Physics, chaired by Fred Scarf, independently questioned whether the radiation problem was insurmountable. With polar orbits designed to minimize radiation exposure and with realistically conceivable improvements in radiation hardening, a mission to mine Jupiter's wealth of planetary information might be possible.

On July 18, 1985, a two-day, NASA-sponsored workshop convened at UCLA to assess the scientific value and the likely feasibility of a Jupiter Polar Orbiter (JPO) mission. The inner jovian system is of interest to a broad range of planetary science; therefore, the workshop participants and contributors represented diverse disciplines.

This report describes two mission designs that the workshop participants considered. These are referred to as the high-periapsis and the low-periapsis options. Both mission scenarios have great scientific value and widespread support. The most important result to surface from the discussions about the orbital requirements is the realization that a well-designed hybrid mission, with reasonable capability to adjust orbit parameters, will very likely be able to satisfy the varied scientific needs to be outlined in the main body of this report.

The report gives the results of preliminary calculations of the radiation hazard for sample missions. The hazard from ring particles is also considered. It presents the workshop's determination of the science that a post-Galileo JPO mission should address and discusses the inner jovian system objects individually. The novelty of the mission concept presents opportunities to use instruments new to planetary exploration and to use conventional instruments in new ways. The report separately explores these possibilities. The final section lists the workshop's conclusions and makes a number of recommendations. The most important of these are given below.

Major Conclusions

- It appears likely that of the two considered mission designs and possible hybrid designs there exist versions that have sufficiently long lifetimes against radiation damage to meet major mission objectives.
- A JPO mission designed to meet a substantial portion of the objectives detailed in this report has very strong scientific justification.
- While either the high-periapsis design, which emphasizes Io and plasma science, or the low-periapsis design, which emphasizes Jupiter science, has sufficient scientific merit to justify a mission, a hybrid design that combines both would yield the greatest return and have far broader support.

Major Recommendations

- NASA should authorize a thorough radiation hazard evaluation in coordination with these recommendations.
- NASA should authorize a parameter study of the two mission designs considered and hybrids of these designs for use in future discussions of mission options.
- In support of the above recommendations, NASA should undertake an advanced study project to determine achievable radiation hardening to set total flux limits for mission designs.
- In view of JPO's likely technical feasibility and its potential for rich scientific yields as assessed by the workshop, NASA should organize a science working team to explore more fully the issues of feasibility and science yields and, if the workshop's preliminary assessment is confirmed, to determine an optimum mission design.

B.2 MISSION DESIGN OPTIONS: ORBITS, LIFETIMES, AND SCIENCE CAPABILITIES

Understanding that the subject mission requires polar orbits to minimize radiation exposure, the workshop considered low-periapsis and high-periapsis mission designs. These are illustrated in Figure B.1. Table B.1 summarizes design parameters relevant to a discussion of each option's merits.

The low-periapsis case (orbit I in Figure B.1) optimizes Jupiter science by providing in situ aeronomy data, high-resolution remote sensing measurements of the atmosphere, and high-order gravitational and magnetic field moments. It allows low-altitude in situ observations of the ionosphere, the thermosphere, the aurora-producing particles, and plasma wave and radio emissions generated under conditions never before sampled in space or in the laboratory. The 11° to 22° apsidal rotation during the estimated nominal mission lifetime, combined with the torus's 14° "daily" wobble, gives adequate torus coverage. The orbit provides off-equatorial, high-altitude passes to observe in situ the waves that scatter the aurora-producing energetic particles.

The rapid latitude sweep of the low-periapsis option precludes a large number of close Io encounters. The low-radiation environment required for visible, ultraviolet, and x-ray wavelength imaging restricts such measurements to an orbit's transpolar segment. While the transpolar segment of the low-altitude option

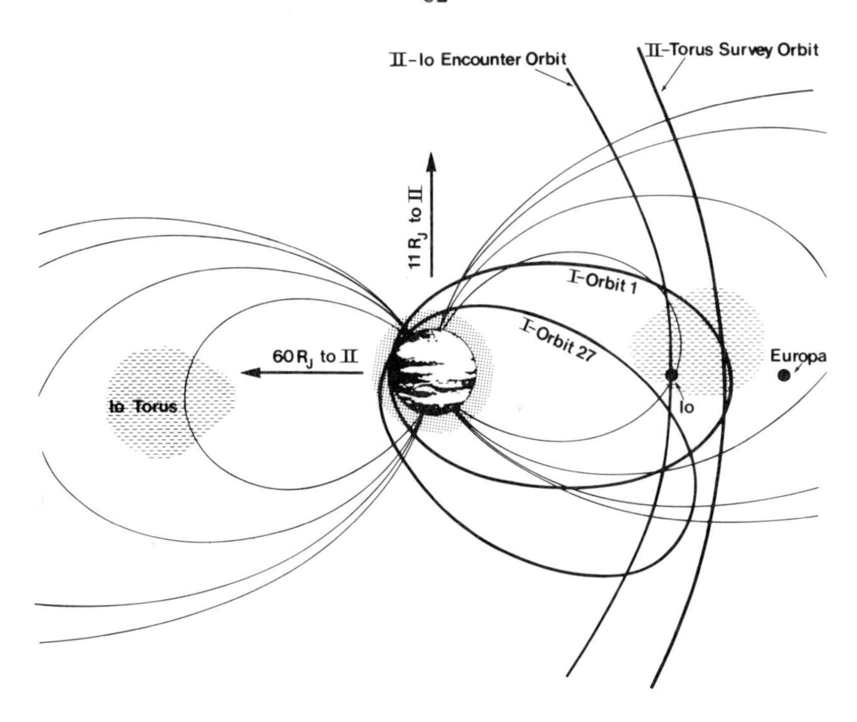

FIGURE B.1 The jovian environment showing the low-periapsis option (I) and the high-periapsis option (II).

aids high-resolution auroral imagery, it limits synoptic aurora and torus studies.

The high-periapsis mission design (orbit II in the figure) optimizes Io and torus science. With relatively small velocity change maneuvers at apoapsis, periapsis can step radially through the torus, giving complete radial and off-equatorial coverage (less than 1 km/s ΔV required). Holding periapsis at Io's orbit permits multiple close Io encounters, enabling in situ atmosphere-aeronomy measurements in Io's leading and trailing, day and night hemispheres. The orbit allows one or more vertical plunges through the northern and southern Alfven wings and Io's wake. The orbit thoroughly probes the energetic particle populations and the off-equatorial region where plasma waves grow. The long nominal mission lifetime permits changing spacecraft and instrument parameters in response to analysis of in-orbit measurements. The high altitude of the transpolar segment allows extensive pole-on

TABLE B.1 Low- and High-Periapsis Mission Design Parameters

	Periapsis	Apoapsis	Altitude at Pole	Period	Number of Orbits to Galileo Dosage (N_G) (preliminary estimate)	Time to Galileo Dosage (T_G)	Line of Apsides Latitude Swing in N_G Orbits
Low Periapsis (Orbit I)	1.014 R_J used as example. Fixed through mission.	8 R_J used as example. Fixed through mission.	Variable 0.3 to 2.0 R_J	1.2 earth days	10 to 20	12 to 24 earth days	11 to 22 degrees
High Periapsis (Orbit II)	5 to 8 R_J used as example. Swept through during mission.	60 R_J used as example. Fixed through mission.	6 R_J (north) 12 R_J	23 earth days	20 to 40	460 to 920 earth days	1 to 2 degrees

visible, ultraviolet, x-ray, and neutral particle synoptic studies of Jupiter's aurora and Io's torus.

The high-periapsis option fails to provide in situ Jupiter aeronomy data, high-resolution Jupiter atmosphere data, high-order magnetic and gravitation field data, and low-altitude auroral particle, radio, and plasma wave data. Remote sensing cannot supply these.

Table B.2 summarizes the main strengths and deficiencies of the two mission designs. While either option by itself could probably be justified in terms of its scientific yield, both omit significant scientific objectives. A follow-on study of a JPO mission should consider a hybrid design that is based on active orbit change maneuvers during the primary phase of the mission.

B.3 SCIENCE ISSUES

Introduction

The workshop participants agreed that while Pioneer and Voyager flybys, Galileo orbiter, and earth-based remote sensing have provided and will provide a wealth of valuable data on the inner jovian system, they leave untouched or unresolved a host of major

TABLE B.2 Comparison of Strengths and Failings of Low- and High-Periapsis Mission Designs

	Low Periapsis (I)	High Periapsis (II)
In situ Io torus	G	G
Off-equatorial plasma waves	G	G
Close Io encounters	F	G
Io-torus coupling	F	G
Torus-Jupiter coupling	G	F
Radiation belts	G	F
"Ring currents"	G	G
Europa	G	G
Amalthea	G	F
Ring	G	F
Low-altitude auroral	G	B
Particles and plasma and radio waves	G	B
Birkeland currents	G	B
Jupiter aeronomy	G	B
High-resolution Jupiter atmosphere	G	B
High-order magnetic field moments	G	B
High-order gravity moments	G	B
Synoptic UV, x-ray aurora and torus	F	G
Imaging fast neutrals	F	G
Reaction time and mission lifetime	F	G

NOTE: G = Good; F = Fair; B = Bad.

first-generation scientific problems. In addition, they have established and will establish a host of important second-generation questions. As a result, a Jupiter Polar Orbiter mission can be designed to address a large, well-focused set of highly significant science issues. Because of the unprecedented variety and the unmatched intensity of the planetary processes in the inner jovian system, the science yield from a comprehensive data-gathering mission to this region could be enormous.

The jovian system may be unique qualitatively as well as quantitatively. According to present thinking its components are remarkably coupled to one another, causing them to behave in many ways like a single interconnected unit. For example, tidal flexing melts parts of Io's interior, forming volcanoes that continually resurface Io with volatile material. These volatiles feed Io's torus by sputtering. From an initial seed population, the torus grows because each torus ion impacting Io sputters much more mass than its loss removes. Centrifugal outflow increases as the torus mass grows until all further addition flows centrifugally outward, and the torus stabilizes. The torus acquires its substance from Io, but it receives its motion from the rotating magnetic field of Jupiter. Jupiter imposes its rotation on the torus with electrical currents, the strength of which is limited by the ionosphere's conductivity. This low-altitude current valve in turn is governed at least partially by aurora-producing particles precipitating from the high-altitude radiation belts. These conductivity-producing particles diffuse inward to their precipitation point by the same electric field that carries torus matter out. Since the outflow is conductivity controlled, this completes a major system-wide feedback loop. After Galileo many of this loop's main components will remain unmeasured and unmapped.

This picture of a microcosm of tightly coupled satellite, torus, energetic magnetospheric particles, and ionospheric plasma on the planet itself has evolved from Goldreich and Lynden-Bell's 1969 unipolar induction model of the Io-Jupiter interaction, as new information was obtained and further research undertaken. As has been noted repeatedly, the interaction between Io and Jupiter is an example of a fundamental type of astrophysical interaction that has a stellar analog in pulsars, accretion disks, and so on. JPO can map and systematically examine the most striking example of this class of interaction in the solar system. The potential to acquire transferable knowledge in this area is great.

A Jupiter Polar Orbiter mission will substantially advance comparative planetology by providing in situ data from the atmospheres and ionospheres of Jupiter and Io and by greatly increasing the coverage and resolution of the imaging and remote sensing of these objects. Supplying these data is certainly one of the remaining major tasks and major challenges confronting NASA.

In this section the post-Galileo science issues relating to the various objects within the inner jovian system are treated separately. The order progresses from Io's and Europa's surfaces inward to Jupiter's interior.

Io and Europa

The two inner Galilean satellites differ markedly from the outer two. Both are essentially rocky objects (Io's density is 3.5 g/cm^3 and Europa's is 3.0 g/cm^3), and both have apparently undergone extensive and complex geological modification.

Spacecraft exploration of these fascinating worlds began with Voyager and will continue with the upcoming Galileo mission. However, major areas of investigation will remain even after the Galileo mission.

Europa's cracked, icy surface was observed only poorly by Voyager, and major questions remain concerning the age of the surface, the nature of the processes causing the tectonic patterns evident in the surface, the extent and depth of the ice layer, and whether liquid water can exist beneath the visible surface. The Galileo orbiter will make several close (<1000 km) flybys of Europa, and will undoubtedly answer, at least partially, many of these questions. There is a high probability, however, that Galileo discoveries about this satellite will lead to many new questions and the need for future investigation.

Io is one of the strangest bodies in the solar system. Its surface is dominated by volcanic activity, including violent, geyser-like eruptions sending material hundreds of kilometers above the surface. The ultimate energy source for this level of activity in such a small object is believed to be tidal dissipation, but current estimates of Io's total energy output are still somewhat higher than theoretical estimates based on tidal theory and orbital evolution. Io has a tenuous atmosphere, probably spatially and temporally variable, including contributions from both an ambient atmosphere and the injection of gases by the volcanoes. Both the atmosphere and the surface are believed to interact strongly with the magnetosphere, supplying about one metric ton per second of oxygen and sulfur to the magnetosphere by poorly understood processes. The Galileo mission will study Io, but has limitations: the orbiter will make only one close (~1000-km minimum altitude) equatorial flyby at the beginning of the mission and will thereafter make synoptic observations from much greater distances. This means that high-resolution imaging and spectroscopy data will be obtained only for a limited region on one hemisphere of Io, which does not contain some of the largest volcanic features observed by Voyager.

To make major contributions to post-Galileo studies of Io, a JPO mission must make several very close encounters with the satellite, at least one within the atmosphere/ionosphere itself. The opportunity to achieve at least one close Europa flyby is also highly desirable. From such flybys, high-resolution visible imaging and surface spectroscopy could greatly extend Galileo's coverage and could make possible study of geological changes covering the period from the Voyager encounters. Multiple close flybys would also allow studies of the higher order gravitational field components, probing the interior and searching for crustal heterogeneities.

Io's Aeronomy

Io's atmosphere is believed to be primarily sulfur dioxide in vapor-pressure equilibrium with deposits on the surface; the pressure may vary from 10^{-8} bar near noon to 10^{-16} bar at night. Of several photochemical products, it is possible that O_2 might build up to a pressure of 10^{-10} bar. All these numbers are uncertain, based as they are on extensive theoretical modeling of two electron density profiles from Pioneer 10 and the ground-based and Voyager identifications of sulfur dioxide frost and vapor. Many of the remarks about Jupiter's upper atmosphere apply to Io as well.

Io's orbit is enveloped in a plasma torus containing atoms and ions of sulfur, oxygen, and sodium. Maintenance of the torus requires a supply of 10^{28} to 10^{29} atoms per second, undoubtedly from Io's atmosphere, which therefore must be completely replaced every few days. The current view is that the atmospheric atoms and molecules are ejected from Io's gravity by the impacts of torus ions co-rotating with Jupiter and therefore passing Io at 55 km/s. This raises the question of what determines the density, because a different density would create a proportionally different source strength. The torus has been stable for several years, but the Pioneer measurements imply a much lower brightness at that earlier time.

Progress in understanding this system requires direct measurement of the atmosphere and its positive ions, such as are readily made by a standard aeronomy package on a spacecraft making a close pass. Several passes at different local solar times would be even better, but even a single one would create a breakthrough in our understanding. Except for a slightly higher mass range for the neutral and ion mass spectrometer, the requirements are identical to those for the similar Jupiter passes (or for many earth missions, Pioneer Venus, and the future Mars Aeronomy Orbiter). Although the torus is discussed elsewhere, its close coupling to the Io atmosphere and ionosphere must be stressed. Its role in atmospheric loss has already been mentioned. It probably also supplies fluxes of electrons that create the ionosphere. Higher energy positive ions from the magnetosphere are believed to play a role in sputtering material from the surface into the atmosphere and from there into the torus.

Io-Torus Coupling

As early as 1965, ground-based observations showed that Io controls radio emissions from Jupiter's ionosphere. Thereafter, theory and observation revealed additional ways in which Io is an unusual satellite. Its orbit is embedded in a dense heavy-ion plasma, localized in a torus ringing Jupiter. The torus plasma, whose source is Io, corotates with Jupiter and consequently sweeps by Io at a relative velocity of more than 50 km/s.

The interaction between a flowing plasma and a conducting body can take many forms, as we know from the diverse ways in which planets and the Moon interact with the solar wind. The nature of the interaction depends on properties of the satellite and of the plasma, and in the case of Io may also depend on properties of the jovian ionosphere. The relevant satellite properties are its height-integrated conductivity, its efficiency in providing neutrals that can be ionized in the nearby plasma, adding "pickup ions" to the flow, and the strength of any intrinsic magnetic field. Further complexity may be added if the satellite is spatially nonuniform (as, for example, if its ionosphere is absent on the dark side). In addition to flow velocity, important plasma properties include density, temperature, and field strength, which, in turn, determine the speed with which perturbations of pressure flow velocity and magnetic field strength and orientation propagate in the interaction region. Jupiter's ionosphere reflects waves back toward the torus with an efficiency determined by its conductivity. The reflected wave returns to the interaction region (thereby modifying the interaction) if the wave travel time is shorter than the time for an element of plasma to flow by Io.

Further discussion focuses principally on the interaction with Io. There the upstream plasma perturbations are small because the flow speed is small compared with wave speeds and no upstream shock is expected. Very near Io, the plasma slows and is diverted, pulling the field with it. Perturbations travel away from Io along the field toward Jupiter's ionosphere, but at the same time, plasma flow sweeps the perturbed region downstream. Thus perturbations are principally found downstream behind a front at angle

$\Theta_A = \pm\tan^{-1}(V/V_A)$, where V is the relative flow velocity and V_A is the Alfven wave velocity. The strongly perturbed region is bounded by "Alfven wings" that extend away from Io toward the northern and southern ionospheres of Jupiter. Details of the plasma and field configurations within the Alfven wings depend critically on the above-noted satellite parameter. Furthermore, as Jupiter rotates, Io's magnetic latitude varies periodically, and it moves up and down through the torus. Both plasma properties and wave transit time to Jupiter's ionosphere vary periodically as a result. Long-term temporal variability may also occur as volcanoes on Io become active or shut off.

Thus, Io's interaction with the plasma in its neighborhood depends on parameters of the satellite ill-constrained by data and on parameters of the ambient plasma that may vary with position and time.

In its single flyby through the region close to Io and downstream in the flow, the Galileo spacecraft will characterize some features of Io. A meaningful upper limit on an intrinsic magnetic field will be obtained. Mass pickup in the vicinity of Io will be characterized, and the mechanisms limiting ionization of neutrals in the vicinity of Io may be defined.

As noted above, the Galileo spacecraft will have only one close flyby of Io. While this will give the first observations of Io's plasma wake and help determine whether Io has an intrinsic magnetic field, we will sample only a small portion of the region of the Io-jovian interactions. For instance, even if Galileo passes through or near the Alfven wing, it will make only a single-point observation. Galileo will not provide any information about the structure of the Alfven wind with latitude. It is important to determine the shape of the Alfven wing by measuring its spatial dependence. For instance, if mass loading is important, the structure of the wake region will provide information about mass pick-up currents. Observations of Alfven waves that have been reflected from Jupiter's ionosphere could give us information about the interaction between Io and the jovian ionosphere. In particular, multiple passes would help to determine how often the reflected Alfven waves can be reflected before they decay.

It is widely believed that neutrals sputtered from Io or its atmosphere are the source of torus plasma. These sputtered neutrals are expected to form a cloud extending well away from Io. Particles in this cloud are ionized to form the plasma torus. Galileo will not be equipped to quantify the properties of this cloud. In particular, we need to know its density and composition. Even if Galileo observations could determine these parameters, we need more than a single pass to observe spatial and temporal variation. Io is thought to be a very nonuniform source of neutrals, and this nonuniform source may be critical for determining the structure of the torus. Multiple observations are required to study the time-varying structure.

The Galileo pass will be downstream of Io. Upstream observations are needed to determine the properties of the incident plasma. In order to make significant advances in our understanding of the controlling processes in Io's environment and the torus, the JPO orbits should be chosen so that multiple passes are made skimming Io's L-shell within the Io torus. Ideally, passes would be made close to Io both upstream and downstream, but with some variation in how far from Io each pass was made. Spacecraft passes through the "remote" (far from Io) torus would be made as well.

In addition to plasma measurements (ion composition, ion temperature, electron density, temperature of thermal plasma, and magnetic field measurements), in situ measurements of neutrals should be attempted in Io's atmosphere. This should be possible with a mass spectrometer at least in Io's coronal exosphere (density of 10^4 cm^3). Remote sensing of ions and neutrals when the spacecraft is far from the torus could also be important.

High-Energy Particles

The inner magnetosphere of Jupiter contains the highest energy, locally accelerated particles in the solar system and the greatest variety of energetic particle source, acceleration, transport, and loss mechanisms within reach of in situ measurements. Io interacts strongly with the energetic particles and leaves its signature on the population in many ways. Including the Io-related processes, many phenomena can only be studied from a close-in jovian orbit. Other processes, which occur at Earth and/or other planets, must also be studied here in order to develop a general theory of magnetospheres that can be extrapolated to the larger scales of astrophysics.

Sources

Energetic particle sources include the solar wind, the upper planetary atmosphere, and ionosphere, and the decay of cosmic-ray albedo products, as at Earth, plus the unique Io torus. The magnetospheric plasma at Jupiter contains the composition signatures of both the Iogenic source (oxygen, sulfur, potassium, and sodium) and the apparent source in the upper ionosphere (H^{2+}, H^{3+}). Because of these "tracer" elements, it is possible to perform unique studies on acceleration and diffusion processes. For example, one would expect that torus ions would diffuse inward, as well as outward, and that it would be possible to separate clearly at L6 the contribution to the inner radiation belt of diffusing ions from Io and that from other sources.

Acceleration

Measurements of the variation of composition and angular distributions with altitude will also provide an exceptional tool in studying acceleration processes taking place in the vicinity of Io's flux tube from those operating at high latitude, above Jupiter's auroral region.

The Voyagers found that the equivalent temperatures of α-particles ($\approx$10 keV) were in the range of >20 to 30 keV at large (20 R_j) L values, and various models have dealt with the heating and acceleration mechanism. If such temperatures persist closer to the planet, and at high latitudes, our current understanding of the heating mechanism would have to be drastically modified. Therefore, it is essential to measure angular distributions of all energetic ions as a function of both latitude and radial distance.

The highest energy particles in the jovian magnetosphere gain their energy by conserving their first two adiabatic invariants while violating the third and moving inward. It is well established that this is a diffusive process, which in many ways resembles Earth's. However, unlike Earth, the diffusion well within Io's orbit is apparently driven by electric fields caused by upper atmospheric tides crossing magnetic field lines. The determination of the diffusion coefficient and its radial dependence is an important objective, both for confirming the driving mechanism and for characterizing the inner radiation belt and its power input. Furthermore, centrifugally driven interchanges and/or large-scale convection are thought to occur near and outside Io's orbit. It is also important to explore the possible role of these processes within the inner radiation belts.

The existence of 1- to 10-MeV electrons in the outer jovian magnetosphere is paradoxical to this model because their energy far exceeds that of particles on the diffusion track. Hypotheses have been proposed for acceleration mechanisms, such as recirculation, direct acceleration by parallel electric field, and magnetic pumping, but there is no consensus and not enough evidence to support one idea or another. As a spacecraft in an inclined orbit crosses the outer magnetospheric field lines at low altitudes, one can seek evidence here, such as field-aligned particle beams, that would reveal where these particles come from.

The results of such a search could lead to an estimate of the lifetime of these electrons, and this, in turn, could lay to rest the question of whether the outer magnetosphere pulses like a clock, as suggested by the Chicago group, or behaves more stably, like a wobbly disk.

Precipitation, Aurora, and the Stably Trapped Limit

In addition to radial diffusion, which accelerates particles, pitch angle diffusion is another mechanism that deserves study. Pitch angle diffusion is caused by the growth of waves that resonate with the trapped particles perturbing their pitch angles. The consequences are radio waves, particle precipitation, and auroras. Besides producing provocative visual and ultraviolet displays, auroras can inject significant amounts of energy to the upper ionosphere and atmosphere, affecting ionospheric conductivity and atmospheric circulation. We know very little about auroras on Jupiter. Although Galileo will probably see visual displays and add vastly to our knowledge in that arena, the actual particle precipitation is better studied from high latitudes and lower altitudes, where the loss cone is large enough to sample. Thus, it is likely that Galileo will increase our appetite for direct measurements of the precipitating particles from a polar-orbiting spacecraft. These measurements should include complete pitch angle distributions and identification of ion species as well as electrons.

If the flux of a trapped species gets high enough, it becomes unstable to the growth of waves, which causes pitch angle scattering and precipitation, which relieve the instability. Both ions and electrons are thought to press this limit between L5 and 15 R_j. The existence of an electron limit at this position is thought to explain the constancy of the decimetric radiation because the inwardly diffusing electrons that eventually produce the synchrotron radiation must pass through the equivalent of a regulator before they get to the site of the radiation. These ideas received some support from previous flybys, but would be greatly enhanced by a more complete survey of electron fluxes in the inner jovian magnetosphere. The principle of a constant flux value is a powerful one for simplifying complex dynamical systems, and one that could have applications elsewhere in the cosmos.

Satellite Sweeping Signatures

The isolated satellite sweeping signatures obtained by past flybys have served as valuable diagnostics of diffusive particle behavior, and obtaining multiple orbital crossings would allow refinement of this very fruitful method. With repeated crossings of the appropriate L-shells, satellite sweeping signatures from the ring, Metis, Adrastea, Amalthea, Thebe, Io, Europa, Ganymede, and possibly Callisto can be investigated as a function of longitude. This would allow a better understanding of microsignatures, and their evolution into azimuthally averaged macrosignatures.

Also to be gained from these sweeping signatures would be information about the gross characteristics of the satellites themselves; e.g., their magnetic field and conductivity. In prior sweeping studies, satellites have always been considered as nonmagnetic, pure insulators exercising a wholly passive role as absorbers of trapped radiation. However, especially in the case of Io, there is ample evidence of more active interactions. Besides the direct manifestations of activity, the absorption cross section depends upon the satellite gross characteristics.

Io and Its Torus

The energetic ion population above 10 keV per nucleon is known to dominate the plasma stress throughout the regions of the magnetosphere so far studied except for the central regions of the Io torus itself. Measurements of energetic ion stresses parallel and perpendicular to the magnetic field are essential to understanding the overall stress balance and configuration of the magnetosphere. Furthermore, energetic ion stresses and their gradients are important for magnetohydrodynamic instabilities such as the interchange and ballooning instabilities. Centrifugally driven interchanges in particular are believed to be the dominant mechanism for transporting sulfur and oxygen ions out of the Io torus.

A low-periapsis orbit would permit exploration of the Io L-shells at lower altitudes. It would even be possible to stay on the same L-shell for an extended length of time and perform detailed energy and angular distributions including their altitude dependences. Either orbit could be used to investigate azimuthal dependences as they relate to distance from Io, and to the tilted and eccentric planetary magnetic field.

Encounters with the Io flux tube itself, and with its Alfven wings, will be possible and should be designed into the mission. This would tell us a lot about particle acceleration, decametric emissions, field-aligned currents, double layers, and so on. Pioneer 11 had a near encounter with the Io flux tube at a high southern latitude, with results that were quite different from those of the Voyager 1 encounter very near Io itself. Repeated encounters at all altitudes would be very desirable for determining the linkage between Io and the planetary atmosphere.

Inner Zone

With the low-periapsis orbit, the motion of the argument of perigee causes the orbit to intersect the equator at all altitudes from perijove to apojove, and also to cross the $L \approx 6$ lines of force at progressively higher latitudes. It would not take an unreasonable spacecraft lifetime to obtain a complete radial and latitude mapping of particle fluxes and behavior. This is the region of the highest energy, locally accelerated particles in the solar system, and it is a unique one for trapped radiation studies, because here synchrotron losses from the energetic electrons enter into the energy balance and particle transport equations. Investigating this region has astrophysical applications, because of radio galaxies and supernova remnants that emit synchrotron radiation. A solid backbone would be provided for many an ethereal theory if we understood in detail the only case that is accessible to *in situ* measurements. All of the mechanisms discussed in this section that pertain to particle acceleration, losses, transport, collective behavior, and electromagnetic emissions can conceivably be found in other worlds.

Radio Astronomy

The Voyager Planetary Radio Astronomy (PRA) experiment revealed that Jupiter has many distinct magnetospheric radio components at frequencies below 40 MHz. The PRA team found that even the decametric wavelength emission (DAM), which had been studied since 1955 by ground-based observers, consists of several independent components. However, PRA was not able to determine if these DAM components were different manifestations of the same source, or entirely different sources emitting in the same general frequency band. No new information on these various DAM components will be provided by either the Galileo or Ulysses radio and plasma wave instruments due to their limited frequency coverage. The Galileo and Ulysses radio instruments will be able to provide some new observations of the long-wavelength extension of the decametric emission (<5 MHz), the hectometric wavelength emission ($\approx$ 1 MHz), but many major questions such as their various source locations or emission mechanisms will not be answered because of the lack of sufficient angular resolution, less than optimal frequency spacing, or lack of simultaneous polarization and direction finding measurements.

Without question, the most outstanding problem in jovian radio astronomy after the Voyager/Galileo/Ulysses missions will be the determination of the source locations of the various radio components. This information combined with measurement of the wave polarization is necessary before true progress can be made in identifying the relationship between the radio observations and those of the other particle and fields instruments. The best estimates from Voyager data and ground-based observations of the locations of the major components of magnetospheric radio emission are only "generic": their specific locations as a function of local time or latitude and longitude are not known.

In addition to the magnetospheric radio sources, radio signals associated with atmospheric lightning (which was observed optically by Voyager) should be detectable by the JPO spacecraft. Neither the Galileo nor the Ulysses radio and plasma wave instruments will make measurements of radio frequency lightning signals because their 5-MHz and 1-MHz respective maximum operating frequencies will be less than the ionospheric cut-off frequency. The Galileo probe lightning instrument will make measurements at low radio frequencies only from below the ionosphere.

Lightning signals were measured at Saturn by the Voyager PRA instrument but not at Jupiter (due primarily to lack of an appropriate instrument operating mode during the flybys). These Saturn observations not only contributed to studies of atmospheric storms, but also led to a new model of Saturn's ionosphere. Since the lightning source was below the ionosphere, the signals received by the Voyager PRA instrument were modified by their propagation through the intervening ionosphere. Analysis of the resulting signal led to the only diurnal profile of Saturn's ionosphere in existence. This same type of analysis at Jupiter could provide an independent means of measuring the ionospheric profile along the sub-spacecraft trajectory.

Finally, data from the Voyager missions have revealed that several components of the jovian radio spectrum are solar wind driven. Thus, a radio astronomy instrument onboard JPO could also serve as a remote monitor of solar wind conditions external to Jupiter's magnetosphere.

Rings

Jupiter's rings contain three major populations of particles: (1) micron-sized particles, (2) much larger particles (micrometeoroid erosion of the surfaces of these "parent bodies" acts as a source to continually replenish the micron-sized particles); and (3) small embedded satellites. Voyager imaging observations discovered the jovian rings and showed that the micron-sized population is present as a very vertically distended "halo" in the inner region of the rings; a major height ring; and an outer, very faint "gossamer" ring. The parent body population is present in the bright ring and gossamer ring region as are the embedded satellites.

The Voyager observations of the jovian rings provided a gross characterization of their spatial location and limited data on the size of the micron-sized particles. The Galileo spacecraft mission will obtain useful additional information about the jovian rings by observing the micron-sized population during solar occultation with the imaging system and near-infrared mapping spectrometer and by observing the other two populations at low phase angles.

These data will provide a 1 to 2 order of magnitude improvement in spatial resolution over the Voyager results.

One major limitation of the Voyager and Galileo observations of the jovian rings is that they were or will be taken very close to ring plane crossing. As a result, it is very difficult to discern the true geometry of vertically distended components, such as the halo component. By observing the jovian rings from positions well out of the ring plane, the JPO mission may be able to better elucidate their true three-dimensional geometry. A related advantage of this mission is that it will obtain data on the plasma and magnetic field environment in the region of the rings. Charging of the micron-sized particles by interactions with the ambient plasma and Lorentz forces acting on the charged grains are believed to play a critical role in determining the vertical thickness of the rings.

Another limitation of existing and anticipated data on the rings is that they have been observed or will be observed only at visible and near-infrared wavelengths. It will be very useful, if possible, to observe the rings at middle and far-infrared wavelengths to further constrain their size and composition. Observations at these longer wavelengths can sense the presence of 100 micron- to centimeter-sized particles, results that are not only of scientific interest, but may be crucial for assessing the hazard to the spacecraft if a hybrid orbit is chosen. Low-resolution spectral data at these wavelengths, if feasible, could provide very valuable data on the composition of all three particle populations by, for example, detecting features associated with the vibrational fundamental of silicates.

Thus observations that may be possible from the JPO mission could make significant contributions toward resolving several fundamental problems concerning the jovian rings:

1. The nature of the three populations of ring particles and their interrelationships.
2. The factors controlling ring structure.
3. The origin of this ring system.

Jovian Aeronomy

The stellar occultation measurements by the Voyager ultraviolet spectrometer provided the first, and still only, quantitative information on the absolute densities of the major constituents in Jupiter's upper atmosphere. These measurements yielded the densities of molecular and atomic hydrogen and a number of important hydrocarbons at pressure levels in the range of 10^{-6} to 10^{-7} bar. The helium abundance was measured only at much higher pressure levels, and thus only extrapolated values are available for the upper atmosphere. The Voyager ultraviolet experiment also provided information on the exospheric neutral gas temperature; this solar occultation measurement gave a temperature of $1100 \pm 200°K$.

It needs to be emphasized that all this information is merely a snapshot of conditions at low latitudes at one particular time. Ground-, Pioneer-, and Voyager-based information (e.g., Lyman emission) clearly indicates that there must be significant temporal, latitude, and maybe longitude variations in the physical processes controlling the behavior of Jupiter's upper atmosphere. Unlike the Earth, Jupiter receives only a very small amount of energy from the Sun; the measured high exospheric temperature implies the presence of some important energy sources (e.g., gravity waves, Joule heating, particle precipitation), but the lack of information on the relevant parameter, such as temperature variations, bulk velocities, conductivities, ion composition, and precipitating particle fluxes, precludes the determination of the dominant process(es).

Each of the two Pioneer and Voyager spacecraft provided two radio occultation profiles of the ionospheric electron densities. This information does indicate that the peak electron densities are likely to be of the order of 10^5 to 10^6 cm^3, consistent with theoretical predictions, but does not allow any meaningful determination of the altitude of the main ionospheric peak or the magnitude of the diurnal and latitudinal variabilities. In situ measurements are required in order to really understand the ionosphere of Jupiter, or even to interpret the radio occultation results. Galileo will not provide any in situ measurements of the thermosphere or ionosphere of Jupiter.

In order to make meaningful advances in our understanding of the physical and chemical processes controlling the behavior of the upper atmosphere and ionosphere, extensive measurements of the composition, structure, dynamics, and energetics of the neutral gas and plasma are necessary. Specifically, neutral temperature and density distributions need to be obtained as functions of height, latitude, and local time. This will require both in situ mass spectrometer measurements of deuterium, hydrogen, and helium at the higher altitudes and remote sensing of these species as well as the hydrocarbons in the lower thermosphere using an ultraviolet spectrometer and/or imaging devices. Neutral wind measurements can be made in situ using a mass spectrometer and/or remotely by Fabry Perot interferometer; such information would be very useful in studying the dynamics and energetics of the thermosphere.

Determination of the properties of the ionospheric thermal plasma is another major area of study that needs to be carried out. Ion composition and temperature can be measured in situ as a function of height, latitude, and local time using ion mass spectrometry. Langmuir probes can be used to measure electron temperatures and densities at high temporal and spatial resolution. Electron density profiles can also be obtained using the radio occultation technique. These measured parameters will help to determine the electrical conductivity of the ionosphere, which must be known in order to understand magnetosphere-ionosphere-atmosphere coupling processes. The ionosphere measurements used in conjunction with the in situ neutral data will permit a consistent picture of the chemistry, energetics, and dynamics of the exosphere/ionosphere to be attempted.

Measurements of the higher energy (nonthermal) particle fluxes are also required, both in the ionosphere and in the inner magnetosphere. The Voyager UVS experiments observed intense Lyman and Werner band emissions at higher latitudes, implying the precipitation of large auroral particle fluxes into the upper atmosphere of Jupiter. Energetic particle precipitation (whether electrons, protons, or heavy ions) undoubtedly drives much of the aeronomy of the thermosphere and ionosphere, even at low latitudes, where airglow observations imply the presence of an unknown energy source. Therefore detailed measurements of the intensity, pitch angle, and energy distribution and composition of these precipitating particles as a function of location and time are of high priority.

Jovian Atmosphere

The Voyager spacecraft encountered Jupiter nearly in the equatorial plane, limiting imaging to the region within 50° of the equator. The Galileo spacecraft encounter will also be nearly in the equatorial plane. In addition, the resolution obtained with the infrared spectrometers strongly limits correlative studies utilizing the visual and infrared data sets. The scale of variations in temperature and composition is too small to be resolved. The Galileo probe will provide just one isolated profile through the atmosphere.

Although little is known of the jovian atmosphere in polar regions, longitudinal coverage at lower latitudes reveals considerable longitudinal homogeneity. The desirable approach is to seek detailed concurrent infrared and visual maps that characterize the abundance of trace atmospheric components and dynamical parameters (temperatures, winds) of the visible cloud deck as a function of latitude. The goal should be to resolve latitudinally as many atmospheric parameters as possible within limited longitudinal regions.

This mission offers new opportunities to obtain the following:

1. Cloud structure—spatial coverage of the cloud deck at 10-km resolution (comparable to a pressure scale height) in visible light.
2. Winds—temporal sampling optimized to observe rapidly evolving eddies at highest resolution and slowly varying, larger areas at lower resolution.
3. Temperature and composition—spatial maps with 100-km resolution at cloud tops and above.
4. Deep structure—microwave sounding of temperature, and abundances of ammonia and water below the cloud deck.
5. Interior structure—increased information concerning gravity harmonics providing constraints for interior models.
6. Chemistry—ratio maps [such as H_2(para)/H_2(ortho), PH_3/H_3PO_4)CO/CH_4, and H_2S/NH_4SH], vertical motions, solar ultraviolet, precipitation, lightning auroras, and chemical processes in the atmosphere.

Fundamental questions that can be addressed concern the following:

1. Poleward heat flux.
2. Large-scale rising and sinking in belts and zones.
3. Deep versus shallow global circulation.
4. Internal heat flux versus latitudinal heat flow.
5. The nature of the chromophores.
6. The abundance of water and hydrogen sulfide.

7. Temperature gradients—vertical, longitudinal, equator-to-pole, belt-to-zone, and so on.

8. Energy balance between long-lived features and their surroundings.

Jupiter's Interior—Magnetic Field

Jupiter, the largest planet in the solar system, is endowed with a powerful and complex magnetic field generated by fluid motion in its interior much like Earth's. Magnetic field measurements sufficient to characterize only the lowest order moments of Jupiter's magnetic field have been obtained by the Pioneer and Voyager spacecraft flybys of the 1970s. These observations demonstrate that while Jupiter's magnetic field is much stronger than Earth's, they are intriguingly similar. Jupiter's dipole is tilted by 9.6° with respect to its rotation axis. For Earth the value is 11.5°. It also appears that Jupiter's dynamo is characterized by the same quadrupole deficit that is evident at Earth. It has been suggested that the quadrupole deficit is a general characteristic of the dynamo process; perhaps the same dynamo dynamics occur in both planets.

Jupiter's dynamo is far more accessible to study than any other planetary dynamo, owing to the large radius of its dynamo generation region. Jupiter's dynamo may occur in a conducting volume bounded by the pressure-induced transition of molecular to metallic hydrogen, which occurs at $\approx$0.8 jovian radii. However, dynamo generation may also occur further out in semiconducting molecular hydrogen. Spacecraft observations of the magnetic field can in the case of Jupiter be obtained relatively close to the core. This is important because the magnetic field rapidly attenuates away from the core ($1/r^{N+2}$, where N is the order), and causes large uncertainty when the measured magnetic field is projected downward onto the core. Thus, detailed and globally distributed observations of Jupiter's magnetic field obtained at close-in radial distances would advance immeasurably our understanding of the dynamo problem.

The low-altitude polar orbit envisioned for JPO can provide a realistic and achievable mapping sequence for the study of Jupiter's magnetic field. Measurement accuracies of $\approx$50 ppm or better, coupled with pointing knowledge of $\approx$10 arcsec, would be sufficient to provide a detailed description of the jovian field, approaching the present knowledge of Earth's field, and more than a thousandfold improvement in current knowledge of Jupiter's low-order moments. For the first time in the century-old endeavor to understand Earth's dynamo, a second example would be available for study.

In addition to providing a vastly improved description of Jupiter's magnetic field, it may be possible to detect the secular variation of Jupiter's field in orbit, should sufficient time be available to complete several mapping cycles. Since Jupiter has strong zonal flows that might extend into the interior and couple to the magnetic field, there may be some rapid secular variation and the possibility of correlating the wind pattern to the field distribution. This would allow the precise determination of the radius of Jupiter's dynamo-generating region, using the frozen flux theorem of Bondi and Gold. At the Earth, application of the frozen flux theorem to the secular variation observations leads to a determined core radius within a few percent of that determined seismically. At Jupiter, this information would determine the pressure and temperature of the molecular to metallic hydrogen transition. In the absence of a detectable secular variation the core radius may possibly be obtained from studies of the spatial harmonic content of the field above the core.

A detailed description of the jovian field is essential to an improved understanding of Io, the torus, and Io-magnetosphere interactions, since Jupiter's ionosphere is an important "valve" in the tightly coupled Io-magnetosphere-ionosphere system. Present knowledge severely limits the accuracy with which the foot of Io's field line can be located in the ionosphere. Only low-order magnetic field harmonics are known at present, and surface magnetic field intensities are known to about 1 gauss. To understand the variations of the torus with longitude and local time, it is first necessary to map the jovian magnetic field to sufficient accuracy. The low-altitude, polar orbit of JPO can provide a satisfactory mapping sequence and detailed knowledge of Jupiter's magnetic field, the coordinate system that organizes the diversity of phenomena of interest in the jovian system. Polar field-aligned currents are an important diagnostic of the jovian magnetosphere and an essential tool in the study of mass and momentum transport in the magnetosphere. The polar field-aligned currents carry anullar momentum from Jupiter to the torus and outward-flowing plasma. The systematic study of polar field-aligned currents in the Earth's magnetosphere by low-altitude polar satellites and ground observatories has been the key to understanding the dynamics of the Earth's magnetosphere and its interaction with the solar wind. Similarly, the systematic study of polar Birkeland currents at Jupiter is essential in monitoring the dynamic interaction between Juipter and Io (Io torus). The low-altitude polar orbit proposed for JPO is ideal for the measurement and synoptic mapping of these polar Birkeland currents.

Jovian Interior: Gravitational Field

The best models of Jupiter's gravitational field were derived through Doppler tracking of Pioneers 10 and 11 during their reconnaissance of the jovian system. Although the two Voyager spacecraft carried improved radio equipment, they did not approach sufficiently close to Jupiter that large improvements in the harmonic coefficients were possible. Currently, J_2 is known to 1 part in 10^5, J_4 is known to 1 percent, J_6 has been marginally detected, and the nonhydrostatic harmonics are consistent with zero. The Galileo orbiter will remain relatively far from Jupiter but may provide small improvements on the field model, indirectly, through precise (optical) measurements of the precession of natural satellite orbits.

The current field model is based primarily on tracking at 1.75 R_j and beyond (since Pioneer 11 was in occultation at its closest approach). As a consequence, the absolute sensitivity in detecting a high-order harmonic, J_{2n}, was degraded by $(1.75)^{-2n}$. For example, the absolute error in J_6 is currently over an order of magnitude greater than the absolute error in J_2. The proposed mission, if it includes even one close encounter (1.05 R_j) will reduce the absolute error in each J_{2n} to an essentially constant value, independent of n. If there are N close encounters (with $N \approx 20$), then the error will be reduced by a further factor of at least $N^{1/2}$. Even more importantly, good spatial coverage will be achieved by multiple close approaches, leading to very tight constraints on nonhydrostatic (especially tesseral) harmonics of the field. The proposed mission should therefore give a value of J_6 to several percent, a detection of J_8, and a detection of any tesseral components exceeding 10^{-7} of the total field.

There are at least three reasons why an improved field model of this accuracy is highly valuable. First, an improved value of J_6 is "determined" by J_4, so the interior models provide rather precise (10 percent error) predictions of J_6. A test of these models is an incremental but nevertheless important step in improving our understanding of giant planet composition and structure, especially of the outermost layer.

Second, the higher harmonics (especially J_6 and higher) are substantially affected by differential rotation, if the observed zonal flow extends deep into the planet. The effect is much larger in Saturn than in Jupiter but may be large enough (at the few percent level in J_6) to be detectable in Jupiter. This is a marginal but potentially very important test of the hypothesis for deep-seated zonal flows.

Third, and potentially most exciting, is the possibility of "seismology"—the detection of tesseral harmonics or time-dependent (i.e., nonsynchronous) contributions to the J_{2n} caused by normal modes of the planet, excited by convection. (Tidal effects are also present.) The likely amplitude of these effects is largely unknown but could be at the level of 1 part in 10^7 and hence marginally detectable. The science of solar seismology is greatly aiding our understanding of the Sun, and the corresponding detection in Jupiter would be a comparable breakthrough for the giant planets. Clearly, it is of importance to establish the capabilities of this mission, especially the effects of such improvements as the use of a k-band signal (32 GHz) in the radio link, which would greatly reduce plasma interference, the use of VLBI techniques to augment conventional Doppler tracking, and the possibility that extremely precise ephemerides of the satellites could further reduce the errors. One possible area of concern is atmospheric drag during closest approach.

B.4 CONCLUSIONS AND RECOMMENDATIONS

Scientific Justification

- A Jupiter Polar Orbiter mission is scientifically exciting for the number of planetary system processes it will elucidate and because of the basic data it will provide for comparative planetology.
- The value to planetary science that measuring Io's activity and associated phenomena adds to a mission to measure Jupiter's

planetary parameters cannot be overemphasized. The combination of low- and high-altitude measurements gives a mission to the inner jovian system extraordinary scientific interest.

Mission Design Feasibility

- Low-periapsis, high-periapsis, and hybrid orbits capable of meeting all major mission objectives are achievable with currently available propulsion systems.
- Through orbit design and radiation hardening, mission lifetimes appear achievable that will enable completion of major mission objectives.
- The ring particle hazard also appears surmountable through orbit design.
- A Mariner Mark II class spacecraft could accommodate the full suite of scientific objectives and engineering requirements, especially power.

Major Measurement Objectives (Moving from Outermost to Innermost)

- At least one but very preferably more than one encounter close enough to Io to enable in situ atmospheric measurements and to determine Io's magnetic characteristics.
- At least one pass through Io's Alfven wings and wake.
- A thorough synoptic map based on in situ measurements of the density contours of the constituents of Io's plasma torus and its thermal structure. (All present maps are theoretically derived from single cuts through the torus—as will be true also after Galileo.) The plasma ribbon separating the cold and warm tori especially needs to be investigated.
- A synoptic map of the amplitudes of plasma waves that scatter the aurora-producing particles.
- Measurements of the species, energy spectra, charge state, and angular distributions of energetic particles to determine acceleration and transport mechanisms, sources and sinks over the energy range 10 keV $< E <$ 1000 MeV per nucleon for ions and 10 keV $< E <$ 50 MeV for electrons.
- Determination of the source locations of the magnetospheric radio components and simultaneous measurement of the wave polarization. Measurement of the radio frequency signals

from atmospheric lightning and use of these signals to determine the peak ionospheric electron density along the sub-JPO track. Remote monitoring of solar wind conditions external to Jupiter's magnetosphere.

- To resolve problems concerning the nature of the three populations of ring particles and their interrelationships; the factors controlling ring structure; and the origin of this ring system.
- Synoptic-scale polar-view images of the Io torus and the jovian aurora.
- Fast neutral particle images of the radiation belts and neutral particle torus.
- Determination of pattern and strength of Birkeland currents at low altitudes.
- Direct measurements of the particle fluxes precipitating into Jupiter's upper atmosphere, in order to establish the mechanism(s) responsible for the observed auroral phenomena.
- Direct measurements of ionospheric flow velocities and field-aligned fluxes to establish whether the ionosphere is a significant source of plasma for the magnetosphere.
- In situ measurements of the thermal ion and electron temperatures to establish the major plasma energy sources and sinks in the ionosphere.
- Remote sensing of the thermospheric winds for a study of upper atmospheric dynamics.
- Synoptic measurements of the thermospheric temperature variations to allow a meaningful study of the energetics of a unique planetary upper atmosphere not controlled directly by solar radiation.
- High-resolution remote sensing atmospheric data.
- Global low-altitude, high-order moment magnetic field data.
- Global low-altitude, high-order moment gravitation field data.

Instrument Requirements

- A full complement of optical imagers, high-resolution interferometer magnetometers, plasma and radio wave detectors, and plasma ion and energetic particle detectors.
- Monochromatic imagers at x-ray, ultraviolet, and visible wavelengths.
- Fast neutral particle imagers.

• In situ aeronomy, ion, electron, and neutral particle analyzers.

Recommendations

• NASA should authorize a thorough radiation hazard evaluation of the considered mission designs, and hybrids of these designs.

• NASA should authorize a parameter study of the two considered mission designs and hybrids of these designs for use in future discussions of mission options.

• NASA should undertake an advanced study project to determine achievable radiation hardening to set total flux limits for mission designs.

• NASA should organize a science working team to explore more fully the issues of feasibility and science yields, and, if the workshop's preliminary assessment is confirmed, to determine an optimum mission design.

APPENDIX: LIST OF PARTICIPANTS

Workshop Organizers

Robert A. Brown, NASA Marshall Space Flight Center
Andrew F. Nagy, University of Michigan
H. Warren Moos, Johns Hopkins University
Fred Scarf, TRW Defense and Space Systems
George L. Siscoe (Host), University of California

Workshop Sponsors

Henry Brinton, NASA Headquarters
William Quaide, NASA Headquarters

JPO Workshop Participants and Post-Workshop Contributors

Abdalla, M., UCLA
Bagenal, F., Imperial College
Barbosa, D., UCLA
Beebe, R., NMSU

Belcher, J., MIT
Brace, L., NASA/GSFC
Broadfoot, A. L., U. of Arizona
Cravens, T., U. of Michigan
Cheng, A., JHU/APL
Connerney, J., NASA/GSFC
Coroniti, F., UCLA
Daiser, M. L., NASA/GSFC
Denner, C., UCLA
Eviatar, A., Tel Aviv/UCLA
Fillius, R. W., UCSD
Frank, L. A., U. of Iowa
Friedlander, A., SAIC
Hinson, D., Stanford
Hsieg, J., U. of Arizona
Hunten, D., U. of Arizona
Ingersoll, A. P., Cal. Tech.
Johnson, T. V., JPL
Khurana, K., UCLA
Kivelson, M., UCLA
Krimigis, T., JHU/APL
Linker, J., UCLA
Luhmann, J., UCLA
Matsumoto, H., Kyoto U.
Moreno, M. A., UCLA
Pollack, J. B., NASA/AMES
Richardson, J., MIT
Russell, C. T., UCLA
Sanchez, E. R., UCLA
Sato, T., TRW
Slavin, J. A., JPL
Stevenson, D. J., Cal. Tech.
Strangeway, R., UCLA
Strobel, D. F., Johns Hopkins U.
Summers, D., Memorial U./UCLA
Thorne, R. M., UCLA
Walker, R., UCLA
Waite, H., NASA/MSFC

Appendix C
Excerpts from the Draft Report of the Workshop on Plasma Physics Research on the Space Station

PLASMA PROCESSES LABORATORY

In 1985 a workshop was held to explore the feasibility of the Plasma Processes Laboratory for the Space Station. Scientists from plasma and fusion research laboratories throughout the United States participated in this workshop.

After three days of vigorous discussion, the workshop participants identified a number of interesting ideas for basic scientific and technological experiments on the Space Station. In each case there is a solid scientific reason for pursuing these concepts on the Space Station as opposed to in the laboratory. Also, plasma physics as a discipline has much to offer the Space Station complex in understanding the plasma environment that surrounds it, and the interaction of a large current-carrying structure (like the Space Station) with this environment. The development of the basic technologies that would enhance the capabilities of future Space Station investigations is also important.

NOTE: These excerpts are taken from "Overview of Space Station Attached Payloads in the Areas of Solar Physics, Solar Terrestrial Physics, and Plasma Processes," by W. T. Roberts, J. Kropp, and W. W. C. Taylor, AIAA Paper 86-2298, September 1986.

The advantages of the Space Station to plasma physics may be categorized into two areas—environmental and operational.

The environmental considerations include:

- The possibility of creating ultrahigh vacuum over a large volume. This may be accomplished by shielding the desired volume from the ambient neutral and plasma flow, creating a high-vacuum wake region.
- An ambient plasma environment uniform over large-scale lengths. This makes it possible to perform experimental studies of processes requiring homogeneous background conditions over interaction lengths attainable only in space.
- The absence of walls and accompanying effects, such as impurity injection, wall currents, and field shorting.
- The large-scale steady plasma flow past the Space Station due to its orbital velocity. This condition is difficult to achieve in the laboratory.
- Combinations of plasma parameters in the Space Station environment that are ideal for qualitative scaling of space phenomena.
- The absence of gravity. This permits a category of experiments that are difficult on Earth, involving colloidal or dusty plasmas as well as certain technology studies involving such effects as breakdown of insulation in mists. Additionally, levitation of components for achieving various boundary conditions or magnetic fields is simplified, possibly leading to previously unattainable field topologies.

Operational considerations include such factors as:

- Long-duration data bases. In contrast to Shuttle-borne missions, it will be possible to explore wider variations of experimental and environmental parameters with correspondingly more comprehensive investigation. Experiments that would yield too few data during a Shuttle flight may be contemplated.
- The ability to modify experiments during the course of an investigation. The scientific return from Space-Station-based experiments can be qualitatively greater because of an investigator's ability to respond to unanticipated results or to modify (to some degree) the experimental configuration as new objectives are indicated by interim data. This mode of operation will lead to a hands-on, laboratory-like capability.

• Maneuverable platforms, tethers, and other adjuncts, which will allow great flexibility in experimental configurations and diagnosis.

• The large-scale sizes available in space, already mentioned above in the context of enabling experiments involving long interaction lengths, will also permit much greater diagnostic access than in ground-based experiments.

The workshop participants, after developing basic evaluation criteria, described nine very broad experiment categories that could effectively be addressed by the Plasma Processes Laboratory.

1. Investigations of the interaction of the large Space Station with the surrounding plasma environment.
2. Investigations of potential buildup on objects in the space plasma environment.
3. Studies of the plasma flow about objects.
4. Investigations of the basic mechanisms of nonlinear particle and wave interactions.
5. Studies of plasma shocks.
6. Investigations of beam-plasma interactions.
7. Investigations of plasma toroids.
8. Studies of the fundamental physics of dusty plasmas.
9. Studies of the physics of plasmas in a microgravity environment.

Appendix D
Probing Fundamental Astrophysical Scales with High-Resolution Observations of the Sun: Prospects for the Twenty-first Century

INTRODUCTION

The past decade has seen a great increase in the sophistication with which we are able to confront the physics of the Sun. Physical theories have progressed from those that assume a simply stratified, equilibrium atmosphere overlying a classical convection zone, to those that recognize intermittent magnetic fields in the convection zone and dynamical structures on all spatial scales throughout the atmosphere. In such situations, it is possible that there is no static equilibrium structure at all. Furthermore, it is believed that all of the observed structures in the Sun, even the largest, are ultimately governed by small-scale processes associated with intermittent magnetic fields or turbulent stresses. For example, electric currents on scales of order 10 km or less may well be the fundamental entity giving rise to coronal heating.

Understanding the physics of the creation and decay of such small-scale currents and their effects on mass and energy transport is thus essential to a proper description of large-scale structures such as coronal active regions, loops, flares, or mass loss in the solar wind.

The interplay between processes occurring on vastly different spatial scales is ubiquitous in astrophysics and heliospheric physics. Small-scale magnetohydrodynamic turbulence is thought to govern the accretion rate of accretion disks feeding compact galactic x-ray sources and black holes in active galactic nuclei; local instabilities such as Kelvin-Helmholtz modes are thought to control the coupling between the magnetospheres of neutron stars and the surrounding matter. Magnetohydrodynamic turbulence is also believed to suppress efficient heat conduction in the hot halos of galaxies and galaxy clusters, thus controlling the rate of accretion in these halos. Further, the same small-scale processes that heat the solar corona are undoubtedly at work in the coronae that are now known to surround many stars.

In all these cases, observations using even the most advanced technology currently conceivable will not allow us to observe directly the controlling small-scale phenomena. In the case of the Sun, however, we can indeed contemplate direct observations. The Sun is therefore a unique tool for understanding a wide range of astrophysical objects, by virtue of the opportunity it affords to observe the underlying physical processes in some detail.

In this study, the workshop participants examine the scientific rationale and technological basis for pushing the study of solar magnetohydrodynamic processes well beyond the regime anticipated from solar space missions planned for the coming decade (~100 km on the Sun, or ~0.1 arcsec at 1 AU). First, the scientific issues involved are addressed, and it is concluded that observations of solar structure on a spatial scale in some cases as small as 1 km would provide an enormous increase of our knowledge of basic astrophysics. Next, practical limits imposed by the emitted flux and the superposition of separate structures along the line of sight are discussed. Finally, the workshop participants address some of the technological challenges to be met if the objectives are to be attainable early in the twenty-first century.

In this study, the workshop participants have adopted as a baseline assumption and prerequisite the existence of a program (broadly speaking, the Advanced Solar Observatory) that will, over the next decade, result in the observational study of the structure and dynamics of the solar atmosphere at a resolution of about 100 km, in visible, ultraviolet, and x-ray wavelengths. Thus, the well-established rationale for observations at these spatial scales is not discussed in any detail.

THE ROLE OF SMALL-SCALE PROCESSES IN THE SOLAR ATMOSPHERE

The solar atmosphere exhibits a vast range of spatial and temporal scales, from coronal eruptions as large as the Sun itself, to flare-related instabilities that involve scales of centimeters and micro-seconds. Moreover, there is mounting evidence that the large and small scales are inseparably linked, such that neither is purely cause or effect; it is the combination that produces what we observe. For example, very small-scale reconnection events in the corona may suffice to destabilize a small portion of a highly stressed magnetic arcade. The dynamic restructuring of the field, and the fluid flows that go with it, may in turn drive further reconnection and further instability, until the entire arcade blows off as a coronal mass ejection. In order to unravel such a synergistic connection between large and small scales, we need to explore the intermediate range of scales.

Some of the basic spatial scales of the solar plasma, such as the ion gyroradius or the Debye length, typically fall in a range (on the order of centimeters) that is not amenable to remote sensing and is unlikely to be explored in the forseeable future. However, important physical processes can be studied on larger scales that are still far below the current limits of observation.

Solar activity and flares are prime examples. Present models of the site of primary energy release and particle acceleration all involve mechanisms—e.g., magnetic reconnection, plasma double layers, strong shock waves, anomalous current dissipation—that require or generate gradients in magnetic field, temperature, pressure, and velocity on scales in the range 0.1 to 10 km. We already know, from indirect arguments, that flare kernels are inhomogeneous on scales of $\leq$100 km. Observations with spatial resolution in the range 1 to 100 km thus offer our best hope for penetrating to the heart of solar flares.

There is now good evidence that much of the flare energy is channeled into the production of beams of nonthermal electrons and that the deposition of such beams in the chromosphere is the principal means for creating the thermal flare plasma. The detection of linearly polarized bremsstrahlung from the impact of an electron beam would constitute a direct signature of this process. Past attempts have been frustrated by the low efficiency of soft x-ray polarimeters and especially by a complete lack of spatial resolution, which dilutes the signal and makes it difficult to disentangle it from instrumental effects. If technological advances make possible an efficient polarimeter with high spatial resolution, it would be possible not only to demonstrate unequivocally the existence of electron beams, but also to trace the beam (or beams) to the site of primary energy release and to flare kernels in the lower atmosphere.

Coronal loops, a fundamental building block of the solar atmosphere, are now recognized to be fundamental to the understanding of stellar coronae and stellar atmospheric activity in general. A decade of intensive work has demonstrated that the gross properties of a loop (e.g., temperature, pressure, magnetic field strength) do not allow us to decide whether it will be stable, let alone how it is heated. We need to know the internal structure of the loop. For example, what are the temperature and density profiles transverse to the major axis of the loop? If the magnetic field is smooth, and if only classical cross-field transport processes are at work, flux tubes separated by only 1 to 10 km can have widely different temperatures and pressures. If cross-field transport is enhanced (due, for example, to a drift-wave instability), the characteristic scale of the gradients may expand to 10 to 100 km. This first of all affects a directly observable quantity, the differential emission measure, but it also bears on the structure of the loop as a whole.

Transport processes in a loop are only one aspect of an even more basic question, the organization of the magnetic field. For example, some theories of coronal heating postulate that intense electric currents flow along coronal loops, and intense currents are associated with small-scale twists in the magnetic field. To study the internal magnetic structure of a loop, either by the use of tracers or by direct measurement, we will require at least 10—and perhaps as many as 100—resolution elements across the minor axis, which extends a few arcseconds.

The structure of magneto-fluid turbulence in a gravitationally bound system is a basic astrophysical problem that can probably be investigated nowhere but in the Sun. Aside from its intrinsic interest, turbulence in the photosphere concentrates and disperses the magnetic field according to the (unknown) power spectrum of the turbulence, and the photospheric magnetic field is the starting point for creating or maintaining fine structure at higher levels in the atmosphere. Existing observations and theory already suggest that most of the magnetic field in the photosphere exists as discrete flux tubes with characteristic diameters of <500 km. Currently planned observations with 100-km spatial resolution should confirm or deny the existence of such flux tubes beyond doubt. However, as in the case of coronal loops, we will almost certainly need to resolve the internal structure of the tubes before we can understand the physical basis for their formation and persistence.

The value of studying small-scale processes in the solar atmosphere can be illustrated by considering the development of our understanding of the large-scale structure and dynamics of the terrestrial magnetosphere, where we already know much about the linkage between large- and small-scale processes. The early observational and theoretical studies of the magnetosphere dealt primarily with large-scale phenomena, and these studies led to a good understanding of the gross, average properties of the magnetosphere. It was found, however, that a sound understanding of the large-scale mass structure, electric current structure, and temporal evolution of the magnetosphere required the detailed study of microscopic plasma processes. Several examples come readily to mind.

The onset of magnetospheric substorms is related to the initiation of fast field-line reconnection in the geomagnetic tail, which depends on microscopic plasma processes operating on spatial scales comparable to the ion gyroradius. Magnetospheric current systems, particularly during substorms, are controlled in part by the field-aligned potential drop that is associated with double layers, anomalous resistivity, and other microscopic plasma processes occurring on small spatial scales. The large-scale mass and energy structure of the plasma sheet in the geomagnetic tail is determined in part by particle acceleration on microscopic scales and by particle precipitation into the ionosphere associated with wave-particle interactions in the magnetosphere. Both the current structure and the particle precipitation problems are intimately related to the large-scale structure and evolution of the aurora.

EXAMPLES ON THE SUN OF SMALL-SCALE PROCESSES OF ASTROPHYSICAL IMPORTANCE

Atmospheric Dynamics and Magnetic Fields on Small Spatial Scales

Several of the space experiments planned for the 1990s have been designed to study dynamical processes and magnetic fields in the photosphere, chromosphere, and corona on scales as small as 100 km. The experiments are expected to provide data that will help to improve our understanding of fundamentally important problems such as (1) the nature of granulation, (2) the interaction between small magnetic elements and photospheric turbulence, (3) the physics of spicules, and (4) the flow of matter and energy between the chromosphere, corona, and solar wind. However, many aspects of these problems will eventually require observations on scales even smaller than 100 km.

For example, the present theory of granular convection envisions a turbulent cascade involving a spectrum of sizes ranging down to scales as small as 100 m. While the smallest scales of the inertial range are probably not observable, it would be important to observe at least one decade (say, 100 km to 10 km) of the inertial range to characterize the spectral distribution and hence to provide a firmer observational basis for the theory of turbulent transport. It is equally probable that such observations will instead demand a physical explanation outside the scope of present theory. The superb white-light observations of granulation obtained by the SOUP instrument on Spacelab 2 have already shown, even with 0.4-arcsec (300 km) resolution, how oversimplified our conception of granular convection has been.

The upward extensions of fine-scale photospheric magnetic fields are associated with the supergranular network, spicules, and other structures that play a role in the transfer of mass and energy between the chromosphere, the corona, and the solar wind. For example, theory indicates that the interaction between photospheric turbulence and magnetic fields on scales of the order of 10 to 100 km may induce electric currents capable of heating the corona. Experiments planned for the 1990s promise to elucidate aspects of these important processes, in particular by observing dynamical phenomena in the cooler parts of the upper atmosphere ($T < 5 \times 10^5$K) on scales as small as 100 km. The results of these experiments will undoubtedly raise new questions regarding the interactions between flows and magnetic fields on scales significantly smaller than 100 km. For the hotter atmosphere ($T > 10^6$K), current observation and theory imply that observations of magnetic field structures and their interactions with flows of matter on scales in the range of 100 to 1000 km are required in order to study the processes involved in coronal heating and the acceleration of the solar wind.

Fine Structure and Dynamics of Flares

There is good observational evidence that the physical processes that control solar flares manifest themselves on scales of <50 km. For optically thin radiation, the observed emission measured from a single resolution element may be combined with an independent estimate of the gas density and the known radiative efficiency to yield the effective volume (V) of emitting material and thereby a conservative estimate ($V^{1/3}$) of the dimensions of the radiating structure. If the structure is filamentary (rather than roughly spherical), its narrow dimension will be smaller than the estimate.

This technique was applied to many extreme ultraviolet (EUV) slit spectra from Skylab. For example, a flare-associated surge was studied in which the Doppler shift of the surge material allowed it to be isolated from other material in the field of view or along the line of sight. Density-sensitive line ratios were used to estimate an electron density greater than 10^{13} cm^{-3} and a characteristic scale of at most 60 km.

The scale lengths derived for this and other flares are consistent with optical spectra and with ultraviolet and x-ray emission measure data from instruments on board the Solar Maximum Mission (SMM). In short, flare structures with characteristic scales extending significantly below 100 km have been inferred at all wavelengths (and temperatures) thus far observed.

The dynamics of flares is linked to their small-scale spatial structure. SMM data have shown that lines such as Ca XIX ($T \sim 10^7$K) show significant nonthermal broadening during the impulsive phase of flares. The degree of broadening is nearly independent of the location of the flare on the solar disk. The broadening may be the result of spatial integration over many unresolved flare loops, of time integration over highly transient, small-scale flows, or of locally isotropic turbulence. It is essential to distinguish between these possibilities by achieving higher spatial (~10 km) and temporal (~1 s) resolution while maintaining spectral resolution.

One of the key advances that emerged from SMM was confirmation of the importance of high-energy electron beams in solar flares. A substantial fraction—and perhaps the majority—of the flare energy may be channeled through beams. When high-energy electrons interact with the ambient plasma, they produce hard x-rays via bremsstrahlung and ultraviolet emission via collisional heating. However, it is unlikely that electron beams are the only mechanism responsible for exciting flare emission at high densities. High-spatial-resolution studies of flares, simultaneously in visible and ultraviolet light, should clarify the role of electron beams and elucidate the other mechanisms. For example, since an electron beam moves along magnetic field lines, a measurement of the area of the ultraviolet surface region in the chromosphere yields an upper limit to the cross-sectional area of the beam at that height (an upper limit because scattering and absorption might produce ultraviolet emission beyond the confines of the beam). Arguments based on the observed relative intensities of ultraviolet and x-ray bursts imply that the diameter of the beams is of order 10 km.

Plasma Heating and Microflares

The fundamental processes leading to the heating of extended stellar atmospheres continue to puzzle us. Previous space experiments have provided observational constraints on some plasma heating mechanisms; for example, OSO-8 data showed that acoustic waves alone cannot heat the transition region and corona. Further progress has been impeded by the fact that the diagnostics necessary to differentiate between competing heating models cannot be applied using present instrumentation: there is simply not enough spatial and temporal resolution.

An example of a basic, unanswered point is whether the heating process is steady or transient. Thus, one possible heating process involves the relief of stresses built up in coronal magnetic fields by the motion of the photospheric footpoint of magnetic field lines, leading to steady flaring over a wide range of flare energies; at the low-energy end of this flare spectrum, the flares are referred to as "microflares." Observational evidence for the existence of such microflares dates back to OSO-7 data of hard x-ray events, and more recently to hard x-ray data obtained with balloon payloads. These data show that the total heating rate by microflares may be comparable to the coronal luminosity if the energy spectrum of electrons responsible for these transients extends down to $\sim$5 keV and the power law connecting the cumulative number of events with 20-keV photon flux above a given threshold extends below the present threshold of 10^{-2} photons/cm^2/s/keV. Thus, the superposition of such events could account for the steady coronal radiative output. Clear observational evidence that heating in solar active regions is largely a transient process would exclude most of the proposed coronal heating processes; the implications for plasma heating in other astrophysical domains (such as accretion disk coronae) would be similarly of major consequence.

An astrophysical situation in which the difference between steady and transient heating is of great current interest is the formation of coronae on very late-type stars (dMe stars) and very young stars (T Tauri stars). For example, high-speed photometry of T Tauri stars has revealed short-term fluctuations similar to solar flares, having power law cumulative spectra similar to those mentioned above for solar hard x-ray transients. Such power law spectra have also been detected in optical observations of ultraviolet Ceti flare stars. Moreover, it has now been shown that there is a good correlation between the mean level of chromospheric and x-ray emission and flare frequency for these low-mass stars. These stellar data support the idea that the observed emission from the hot outer layers of late-type stars is the result of a temporal superposition of transient energy release events. Thus, a very important question is whether quiescent EUV and x-ray emission from solar active regions (and even the quiet corona) is due entirely to superimposed transients.

In order to answer this question, one is faced with the difficulty that, as the energy of individual events decreases, their visibility above background at any given angular resolution decreases as well. This problem is not acute at high photon energies (>20 keV) because the slowly varying (background) component at these energies is unimportant even for full-disk (completely unresolved) observations, at least at current sensitivity levels. However, at lower photon energies (<5 keV) the slowly varying component begins to affect our ability to detect individual events, so that imaging becomes essential.

Since the observed duration of the microflares at high photon energies is a few seconds, it also becomes necessary to consider time-resolution constraints. For example, because of the thermal inertia and the cooling rate of the transiently heated gas, it is difficult to detect low-level microflaring at soft energies because the coronal gas cooling time (10^3 s) far exceeds the time scale of the transient heating event itself. In contrast, ultraviolet and EUV emission from gas at lower temperatures ($\sim$1 to 3 $\times$ 10^5K) can vary at the transient time scale because the corresponding cooling times are sufficiently small. However, the smallest events will require high spatial resolution in order to avoid averaging within a resolution element.

An estimate of the size of the elemental ultraviolet-emitting transient gas at the lower end of the energy spectrum is given by the estimated extent of flare kernels, <60 km (see above). It is necessary to extend observations to even smaller scales in order to resolve these events and to establish the cumulative contribution of the power law microflare spectrum to the total ultraviolet/EUV luminosity of active regions.

OBSERVATIONAL CONSIDERATIONS

The above discussion has shown that much would be gained by observing physical conditions in the solar atmosphere on spatial scales considerably finer than the best anticipated from current programs. Later in this appendix the technological considerations that enter into observing small-scale structure are discussed. However, intrinsic limitations that arise from the available photon flux and the geometry of the source are first discussed.

The surface brightness of a feature limits the flux that can be collected from a single resolution element in a time shorter than the characteristic evolution time scale of the element. This time scale could be as short as the sound or the Alfven crossing time—often one second or less for subarcsecond structures.

At what rate will photons arrive at 1 AU from an element of EUV-emitting material? If the element is a cube of side d, the number of photons detected per second at 1 AU will be

$$1.0(n_e^2\Lambda_{\rm max}/10^{-2})(d/10\,{\rm km})^3(\lambda/100{\rm \AA})(D/1\,{\rm m})^2(\epsilon/10^{-2}){\rm photon\ s}^{-1}$$

where $\Lambda_{\rm max}$ is the cooling coefficient, $n_e{}^2\Lambda_{\rm max}$ is the volumetric radiative power, D is the aperture of the telescope, and ϵ is the overall efficiency of the instrument (photons counted divided by photons incident). It is assumed that the passband $\Delta\lambda$ at wavelength λ includes all the radiation from the line or lines, and that there is one element in the field of view. If the telescope resolves distances smaller than d, the counts are assumed to be integrated over $d \times d$.

In the visual and near ultraviolet, the expected flux is encouraging for a meter-class telescope. For example, the C IV doublet near 1548 Å produces $n_e{}^2\Lambda_{\rm max} \simeq 1.0$ at a pressure $n_e T = 10^{16}$

(roughly appropriate for an active region loop), leading to a photon flux $\sim 10^3$ photon/s from a 10-km element. However, for higher temperature lines ($T > 10^{5.5}$) in the EUV, typically $n_e^2 \Lambda_{max} < 10^{-2}$ at the same pressure, and, at 1 AU, only a few photons would be detected per second. There could be little confidence that the element would not evolve substantially or even disappear during the time necessary to build up an image with adequate signal-to-noise ratio.

These figures refer to structures of roughly average brightness. In flares, pressures 10 or more times greater have been inferred. The density enhancement in microflares is not known, but could be comparable. Thus, there is good reason to expect that some structures, some of the time, will have 100 or more times the average brightness.

Nevertheless, if we wish to observe typical EUV structures of a size of <10 km on the quiet Sun, there is a flux problem. This could be addressed either by larger collecting areas (of order 100 m^2) or by meter-class telescopes in near-Sun orbit (say, at a distance of 0.1 AU).

The example given above treats a single emitting volume 10 km on a side. Some lines of sight may include several structures. Although the flux at Earth increases accordingly, the flux problem is alleviated only at the expense of source confusion, since the individual structures may have distinct physical conditions. Source confusion could be addressed by stereoscopic observations, in which two or more spacecraft simultaneously examine the same region of the solar surface from different vantage points.

A third problem is the reduction in contrast of very fine structures embedded in an optically thick medium such as the photosphere. In this case, image sharpness is limited by the photon mean-free path (~100 km in the photosphere for visible light). This problem is less important for observations of Doppler or Zeeman line shifts to the extent that scattering causes diffusion in position but not in wavelength. Moreover, the problem disappears for optically thin media such as the chromosphere (for many spectral lines), transition zone, and corona.

It may be that the twin problems of flux limitation and source confusion will, more than purely technical considerations, limit the remote sensing of solar features to scales of 1 km.

HIGH-RESOLUTION IMAGING IN THE EUV

Because of the absence of atmospheric distortions and the accessibility of all wavelengths, high-spatial-resolution observations of the Sun are most advantageously carried out from space. However, in the past, few space-based telescopes approached their theoretical resolution. Instead, resolution was limited by focal length restrictions, the quality of the optical systems, and the properties of the detector. For example, the High Resolution Telescope-Spectrograph (HRTS) ultraviolet telescope is limited to 1 arcsec (rather than 0.1 arcsec) because of overall size and detector limitations. The Solar Optical Universal Polarimeter (SOUP) visible-light telescope (30-cm aperture) is so far the only telescope to operate in space (Spacelab 2, 1985) at its theoretical resolution of 0.4 arcsec. The projected resolution of the Hubble Space Telescope (HST) approaches its diffraction limit of 0.05 arcsec in the visible and near ultraviolet, but not in the ultraviolet. The 1-m optical telescope of the High Resolution Solar Observatory (HRSO) will have a diffraction-limited resolution of 0.13 arcsec across the visible spectrum.

Thus, in the visible wavelength region, small space-based telescopes have achieved their diffraction limit, and large telescopes will soon do so. If the diffraction limit could be achieved at EUV or soft x-ray wavelengths, we would gain an order of magnitude or more in angular resolution. This appears feasible if progress is made in several areas:

1. Mirrors of higher figure quality must be polished, corrected, or actively controlled.

2. At wavelengths below 1000 Å, nearly diffraction limited resolution probably requires normal incidence optics, and this in turn entails the development of multilayer, high-reflectance coatings. The use of phase surface correction and adaptive optics techniques already explored in the visible may also be required.

3. The linear resolution of available detectors (film or two-dimensional solid state) is inadequate. For example, a diffraction-limited 1-m telescope would require an $f/3000$ focal ratio in order to match a detector with 10-μm pixels. Therefore, without enlarging optics (see the subsection below on detectors), detectors that have submicron resolution elements must be developed.

Although high spectral resolution does not appear in itself to pose technical difficulties at this stage, a combination of high spatial, spectral, and temporal resolution may strain the photon flux limits discussed in the section above on observational considerations.

Interferometry is another technique for achieving extremely high resolution and is considered below. Also addressed below are the problems of pointing systems and the question of where in space these instruments would best be operated.

The workshop participants conclude that, in all the abovementioned technical areas, rapid developments are being made today and that ultrahigh-resolution telescopes or interferometers will be feasible within the time frame 1995 to 2015.

Detectors

Currently available detectors (including charge coupled devices (CCDs), television-type systems, and electronographic cameras) are capable of pixel size near 10 μm and a format of about 2000 $\times$ 2000 resolution elements. TV-type systems have flown on many missions, and recent improvements in these older technologies have shown that 2-μm pixels and 4000 $\times$ 4000 format can now be achieved, while maintaining high quantum efficiency throughout the EUV and soft x-ray region. It appears likely that a pixel size of less that 1 μm can be achieved with very fine grain phosphors or thin layers of evaporated scintillator applied to the aplanatic focus of a high-quality lens (such as a Burch objective) coupled to a high-resolution TV or TV-type detector. Even higher resolution, of order 0.2 μm, can be achieved with a semitransparent photocathode followed by reimaging of the photoelectrons onto a suitable detector, with acceleration of the electrons and expansion of the plate scale achieved simultaneously.

Some increase in plate scale can be accomplished by suitable design of the optical system, such as using relay optics in a grazing incidence configuration or using a Cassegrain design with normal incidence optics and x-ray multilayers. However, at the highest magnifications, the mechanical positional and alignment tolerances will make such telescopes very expensive. In order to achieve the highest resolution in an instrument of practical size and cost, it will be necessary to build detectors with micron-size pixels. For solar observations, large image format is also needed, which can be obtained with a very high resolution television tube, a large format CCD or MAMA (multi-anode microchannel array), or by a mosaic of detectors. None of these techniques has yet been shown to work in a large enough format for high-resolution solar applications, and only systems that enlarge the focal plane image onto the detector have the required pixel size.

Multilayer Coatings for Normal Incidence Optics

Recent progress in techniques for the deposition of high-quality, thin-film multilayers onto optical surfaces allows us to begin planning a new generation of instrumentation for the EUV and soft x-ray region. We can now consider instruments that are qualitatively different from those traditionally employed in this wavelength region—ones that will have the spatial and spectral characteristics of the highest quality ground-based optics and some spaceborne instruments such as the HST and the HRSO.

High-Resolution Imaging

All conventional coatings have low normal-incidence reflectivity in the EUV. This imposes severe constraints on EUV instrument designs for space astronomy. Below 400 Å, normal incidence designs have led to unacceptably low throughput, especially in applications involving multiple reflections. However, further development of a new technology—multilayer, thin film coatings that have high reflectivity at normal incidence at wavelengths down to 20 Å—will allow the use of EUV instrumentation that incorporates normal incidence components. In addition, such coatings can be applied to glancing incidence mirrors in order to enhance their short-wavelength performance.

The principal advantage of normal incidence optics in comparison with glancing incidence is the ability to achieve substantially higher image quality with a given level of effort. In addition to this simplified manufacture and testing, normal incidence telescopes will generally be lighter and more compact, and can provide high-resolution images over a broader field.

Multilayer coatings increase x-ray reflectivity by exploiting the fact that a discontinuity in the complex refractive index causes reflection of an incident wave at the boundary. In the EUV, this reflection is small because the material has a refractive index near unity; however, by adding more boundaries such that all reflections add in phase, the reflected intensity increases as the square of the number of boundaries N. If spacer layers with zero absorption could be found, 100 percent reflectivity could be achieved. In practice, N—and therefore the maximum reflectivity—is limited by absorption in the spacer layers.

To achieve multilayer coatings with high reflectivities, uniformly thick layers and sharp, smooth interfaces are required. These requirements severely limit multilayer performance at short wavelengths. It appears that the fundamental limit is the diffuse nature of the atoms themselves, which limits us to wavelengths longer than about 10 to 20 Å. In theory, reflectivities of 30 to 50 percent should be obtainable throughout the 20- to 300-Å range, and such values have already been achieved at the longer wavelengths.

In general the substrate must be smooth to better than one-tenth of the multilayer period, because any roughness will be replicated in the multilayer and degrade its performance. Such surface quality appears to be within the state of the art, particularly on small substrates, even in the 20-Å wavelength region for which 1-Å root-mean-square surface smoothness is required. The deposition of uniform layers onto meter-class substrates should be readily achievable by methods similar to those used for longer wavelength optical coatings.

Spectroscopy and Spectroscopic Imaging

By an effect analogous to Bragg reflection, normal incidence multilayer optics provides an inherent spectral filtering that allows simultaneous high-angular-resolution imaging and spectroscopy. The multilayer spectral resolution $\lambda/\Delta\lambda$ is roughly equal to the number of layer pairs and varies from $\sim$10 at 200 Å to $\sim 10^3$ at 20 Å. Imaging in a narrow bandpass that reflects a single line, or lines, of a single ionization state, formed over a narrow temperature range, may provide the best opportunity to study fine-scale solar features at EUV wavelengths. In such nearly isothermal images, structures are more sharply defined and are more readily interpreted than images averaged over a broad temperature range.

Regions of the solar spectrum have been identified that contain several emission lines of comparable strength spanning the 10^5K to 10^7K temperature range for which multilayers with high-reflectivity can be produced. In such cases, dispersive imaging (similar to that obtained in the EUV on Skylab) will yield useful scientific diagnostics. To accomplish this, a transmission grating introduced into the optical path of a normal incidence telescope produces a high-resolution image in each of the emission lines.

Multilayer coatings have been applied to reflection gratings in order to enhance their efficiencies in the EUV by about a factor of 3, demonstrating the feasibility of their use in spectrographs. A simple focal plane, stigmatic spectrometer using a normal incidence, toroidal grating can be constructed with a moderate size mirror acting as a feed to the grating. Resolution of $\lambda/\Delta\lambda \sim 3 \times 10^4$ can be achieved, thereby permitting separation of nearby spectral lines, measurement of line broadenings and asymmetries, and determination of radial velocities to a few kilometers per second.

Glancing Incidence Optics

Although the multilayer technology just discussed opens up the exciting prospect of applying the fine-scale imaging capabilities of normal incidence optical systems to solar observations at soft x-ray and EUV wavelengths (30 to 400 Å), there will still be a need for glancing incidence optical systems at shorter wavelengths. At present, the inherent size of the atoms used to manufacture multilayers (W, Ir, C, Ag, Si, Mo, and so on) appears to restrict their use to wavelengths above about 30 Å. In addition, even though substantial improvement in the pixel size of photoelectric detectors is expected in the very near term, it has already been noted that the pixel sizes required for milliarcsec imaging at these shorter wavelengths become remarkably small. One way to handle this problem is to use multielement telescope systems with substantial magnification. However, the magnifying elements in these systems are then subject to high concentrations of solar flux and the multilayer coatings on them are subject to photopolymerization of hydrocarbon contaminants. This buildup of molecular contamination on the mirror surfaces not only decreases the reflectivity in the bandpass of interest (somewhere between 30 and 400 Å) but also increases the absorptance of the intense visible light flux falling on them (about 50 percent of this flux is absorbed by a clean, uncontaminated multilayer surface) and could lead to thermal runaway and destruction of the coatings.

Glancing incidence optical systems, at $\lambda < 30$ Å, are insensitive to such contamination and can be used to image a very broad wavelength band, including wavelengths below 30 Å. However, just as multilayer performance is limited by the surface roughness of their substrates, the performance of glancing incidence optics at 200 to 400 Å is currently limited by "mid-frequency" surface ripple with a spatial scale of a few millimeters. Until now, the optical transfer function of glancing incidence telescopes has been characterized by very narrow cores and broad scattering wings that contain the great bulk of the energy in the point source image.

Recently, computer-controlled polishing technology has been applied to the production of glancing incidence optics, and the surface smoothness achieved indicates that good imaging at the arcsec level can now be achieved. The amplitude of the mid-frequency ripple on these surfaces is about 300 to 350 Å root-mean-square, and within the next few decades it should be possible to produce surfaces with mid-frequency ripple of 30 Å root-mean-square or better. This improvement in surface smoothness by an order of magnitude will reflect itself in an improvement in attainable resolution by an order of magnitude, the limiting factors then being the ability to achieve and maintain relative alignment of the primary and secondary mirrors and the diffraction effects caused by the large central obscurations of glancing incidence telescopes.

Primary Figure Correction and Control

The primary resolution of the telescope is affected by the alignment accuracy of the optical train and the consequences of variations in element position having temporal periods of hours (thermal) to tenths of a second (vibrational). In addition, there exist residual figure errors on the primary and subsequent mirrors, and smaller-scale errors induced by thermal loading, polishing imperfections, and phase distortion due to residual coating nonuniformities (which may be wavelength-dependent).

The goal for reduction and compensation of these errors will be determined by the desired linear resolution, the incident flux available, and engineering feasibility, not necessarily the diffraction limit at all wavelengths. The diffraction limit at 1-m aperture and $\lambda = 1000$ Å (~0.02 arcsec) may be realistic, while for $\lambda = 30$ Å (~0.007 arcsec), it may not.

Correction may be accomplished by a hierarchy of techniques. Correction of constant residual figure errors can be done to potentially high precision by fabrication of a fixed compensation plate generated by photoresistive techniques. Because multilayer dielectric coatings conform very accurately to the substrate on which they are deposited, resolution is critically dependent on substrate figure. After polishing, current meter-class mirrors show residual deviations of 300 Å peak-to-peak, which would severely degrade short-wavelength resolution. Because the defects are fixed, a corrector with conjugate phase errors can be employed to compensate the errors in a subsequent pupil plane. The completed instrument can be used as one arm of an interferometer, with an accurate reference surface in the other arm and the photoresist coated corrector in the fringe plane. For small enough residuals (<800 Å), the exposure creates the appropriate phase conjugate, and subsequent development of a multilayer coating of the corrector completes the process. Current experiments suggest that residual surface errors of 10 Å may be achievable.

Time-varying figure errors of low spatial frequency (generally thermally induced) can be measured by wavefront sensing or interferometric techniques and used to correct the primary mirror figure (by means of electromechanical actuation), perhaps supplemented by an active mirror operating in a subsequent conjugate plane. The technologies of wave front sensing have already shown accuracies of better than $\lambda/500$ in the visible. Many forms of active or adaptive corrector mirrors employing piezoelectric, electrostatic, and other principles have been built, and high-performance stabilization of large space mirrors is currently under vigorous development. Furthermore, wave front sensors can easily resolve 10,000 or more pupil elements, which can be used to drive lower resolution correctors. The residual error is a dynamic measure of the instrumental optical transfer function, which can be used for post facto resolution enhancement (see following subsection).

These technologies are applicable to both normal and glancing incidence configurations, and they are already sufficiently mature to include in the basic design philosophy of any instrument.

Resolution Enhancement Techniques

The angular resolution of the proposed optical systems may not in practice be diffraction limited at EUV wavelengths, even with correction and control. Several techniques can be applied to improve the spatial resolution after detection, with the degree of improvement dependent on wavelength, flux levels, and feature contrast. All of these techniques should yield some improvement even at very short wavelengths.

Aperture Synthesis (AS)

In AS, a mask with a set of transmitting holes is inserted across the primary mirror, or in a reduced image of the primary. The resultant subapertures are positioned so that their pairwise autocorrelations sample the (u,v) (spatial frequency) plane to the greatest degree possible. Conventional images are recorded with the array in place. Subsequent analysis of these coded images allows estimation and correction of the phase errors across the subaperture, and eventually, reconstruction of enhanced images.

The major advantage of this technique is its simplicity of implementation, with full image reconstruction from one or two exposures. Its drawback is that perhaps only 10 to 20 percent of the total flux is transmitted by the masks.

Speckle Imaging

Speckle imaging techniques, which have been used to recover diffraction-limited images with large-aperture telescopes through atmospheric turbulence, may be used to overcome the effects of the uncorrectable errors in EUV optics. If pairs of small apertures are scanned across an optical pupil plane, and images are recorded for each position of the subapertures, then averaging of the complex autocorrelations of the images would allow high-resolution recovery. Each position of the pair of subapertures provides a statistically independent realization analogous to random atmospheric fluctuations between short exposure frames in ground-based speckle. The major problem is that the time required for this serial procedure is significantly longer than a single-frame technique such as AS, so its time resolution is limited. Its advantage is the potential to correct to much finer scales.

Deconvolution

A by-product of an actively controlled optical system is that an accurate measure of the optical transfer function is obtained by the system wavefront sensor. Because the sensor uses its own source, good signal-to-noise and accuracies better than $\lambda/1000$ (visible) can be achieved. An accurate deconvolution of each image with the system point spread function may therefore be applied as a post-processing step. Only the presence of zeroes in the optical transfer function limits this technique, and these may be eliminated by adequate compensation with active optics.

INTERFEROMETRY

One way to obtain very high angular resolution is through spatial interferometers, using either arrays of telescopes or arrays of flat mirrors feeding a central beam-combining telescope. The goal of such an interferometer would be very high angular resolution imaging (milliarcseconds) from visible wavelengths (7000 Å) to the EUV (~500 Å). Although the precise form of such an interferometer remains to be defined, at least two options are available. Both are so-called monolithic arrays, which means that the interferometer system is mechanically coupled in such a way that the pathlengths between the Sun and the focal plane are approximately equal in all legs of the interferometer. Both are also designed to have a wide field-of-view, which means that the pupil configuration at the entrance and exit of the interferometer is preserved.

In the first, a "thin monolithic interferometric array," HRSO is used as a beam-combiner. As in the early Michelson and Pease interferometer, pairs of flat mirrors are used to feed the light into the telescope from points outside the telescope. In one version, five of these 40-cm-diameter interferometer beams are arranged in a pentagon with a 12.5-m diameter, feeding the 1.0-m diameter optical telescope, giving 10-milliarcsec resolution in the visible or 3-milliarcsec resolution in the C IV lines near 1550 Å. It could be assembled in space to the required dimension. Because of the large coherence lengths ($\lambda^2/\Delta \sim 1$/mm) that characterize high-spectral-resolution solar observations, the tolerances on the lengths of the interferometer legs are rather loose for narrow bandpass observations. Such an interferometer could be implemented at an early date at a relatively low cost. For imaging, it will be necessary to determine the relative pathlengths of the interferometer legs. This requires the development of the equivalent of phase closure techniques used in radio image synthesis telescopes like the VLA but for solar observations, and image synthesis by rotation of the array and by changing the length of the arms. Because of the flux limitations in the ultraviolet, phase closure has to be done in the visible wavelengths.

The second type of interferometer is a "thick monolithic array" in which the interferometer elements are coupled by an optical metering system to achieve cophasing. It gives complete (u,v) plane coverage in one dimension. Rotation of the array will fill the (u,v) plane. Phase closure is not needed in this case.

Both types of arrays are best used in conjunction with HRSO on the Space Station or on co-orbiting platforms. Before the implementation of a free-flying solar space interferometer, it is important to determine the visibility of different scales of solar fine structures by means of two-element interferometers, using ground-based interferometers for the visible and, e.g., a EURECA-class experiment for the far ultraviolet, allowing milliarcsec resolution. Phase closure techniques for solar observations can also be pioneered from the ground in the visible region of the solar spectrum.

POINTING SYSTEMS

The SOUP telescope on Spacelab 2 has demonstrated that it is possible to point a white light solar telescope with internal image stabilization to a stability of 3 milliarcsec using a 50-Hz control bandwidth. Similar image control techniques have locked two telescopes to an accuracy of a few milliarcseconds, which is essential for ultraviolet observations because it is often necessary to use detectors that are insensitive to visible light or require that visible light be excluded from the telescope.

The new generation of solar experiments should provide for multiple-aspect ("stereo") studies of solar structures. Since the structures to be studied may be in the 10-milliarcsec size range, it would be desirable to be able to point several widely separated telescopes at the same feature on the Sun with an absolute accuracy of a few milliarcseconds. Since the average solar diameter is known to be stable at that level or better for periods longer than a month, a pointing system based on metrology of the average solar limb should yield the required accuracy.

If the solar diameter and the pointing position are measured with a laser interferometer, which should yield $\lambda/10$ accuracy, the location on a 5-cm solar image can be measured to 1.6 parts in 10^6. Since the solar diameter is about 2000 arcsec, this represents a positional accuracy of 1.3 milliarcsec. Co-pointing of two independent telescope systems also requires that their image planes be homogeneous to this level and that their roll orientation be measured to 0.25 arcsec.

SITING

From the preceding discussion, it is clear that spatial resolutions substantially below 0.1 arcsec, at wavelengths from 30 Å to visible light, can be achieved with techniques that will be perfected in the twentieth century. The question of where to situate the resulting high-resolution solar instruments must now be addressed.

Near-Sun Orbit

Close approach to the Sun offers a conceptually simple way of achieving ultra-high angular resolution with instrumentation of modest aperture, and it may be the only means of obtaining 10-km resolution of the solar surface with sufficient EUV flux. The basic technology required for near-Sun orbits (e.g., heat shielding, telemetry through the corona, injection into the correct orbit) will be developed for the Starprobe program. It would seem profitable to use this technology in follow-on missions carrying solar imaging experiments. Some of the additional studies for shielding the instruments were done in the preliminary stages of the Starprobe program and indicated that this concept is quite feasible.

Heliosynchronous Orbit

A spacecraft at ~30 solar radii—approximately 0.1 AU—will have a 25-day orbit around the Sun, and in the ecliptic plane will hover over a fixed point on the Sun. In addition to important long-term studies of the evolution of solar structures, and stereoscopic measurements of fine features when combined with another spacecraft in similar or near-Earth orbit, such a vantage point offers an increase in spatial resolution by a factor of 10 and flux by a factor of 100 compared to an orbit at 1 AU.

Lunar Basing

A solar observatory on the Moon would operate under almost

ideal conditions, on a highly stable platform, free from atmospheric turbulence and contamination, and able to view the Sun continuously for 14-day periods. Such an observatory could operate either manned or unmanned, and should be considered in the context of any future Moon landings. Two lunar observatories separated by 180° would give nearly 100 percent coverage.

Solar Orbit at 1 AU

Stereoscopic observations of the Sun can be achieved with two observatories in orbit at 1 AU widely separated ahead of and behind the Earth. Alternatively, one of the observatories could be placed at a Lagrangian point. It is felt that the simultaneous pointing accuracies needed to achieve 0.01-arcsec resolution can be achieved.

Earth Orbit—Free Flyer

The ideal near-Earth orbit for a solar observatory would be a high-inclination, Sun-synchronous orbit, i.e., one perpendicular to the Earth-Sun line. The first mission of the HRSO (High Resolution Solar Observatory) on the Space Shuttle is planned for a high-inclination orbit, which will allow long periods of uninterrupted observation.

Manned Vehicles—Space Shuttle

It is expected that the Space Shuttle in some version will still be flying at the century's end. Spacelab has proven itself a valuable vehicle for astronomical observations. Advantages of Spacelab include significant real-time scientific interaction with the experiment as well as the ability to use either solid-state detectors or recoverable film. While solar observations can make effective use of week-long flights, longer missions could significantly increase the scientific return. Disadvantages include relatively short observing periods for low-inclination orbits and the potential for contamination of highly sensitive ultraviolet and EUV normal-incidence optics. The principal contaminant is expected to be thin (10 to 20 Å) layers of polymerized hydrocarbons, which are opaque to light between 300 and 1300 Å. Solar telescopes are particularly prone to organic contamination because the solar ultraviolet light polymerizes hydrocarbons rapidly. Any site in space where such telescopes are deployed requires rigid material selection standards if the telescope is to operate for an extended period. Skylab experience demonstrates that this can be done.

Space Station

The Space Station shares many of the advantages and disadvantages of the Shuttle. An additional concern is the expected broader spectrum of vibrations and disturbances, which make pointing difficult; isolation platforms will probably be required. Contamination is also of great concern, and it is imperative to pay careful attention to materials selection for the Space Station from the outset. Of course, a co-orbiting platform circumvents many of these difficulties and would provide a very attractive site for solar observations.

PROSPECTS AND LIMITATIONS OF GROUND-BASED OBSERVATIONS

Radio Interferometry

The usefulness of high-angular-resolution observations for solar physics has been demonstrated at radio wavelengths since the early 1970s. The use of large-synthesis radiotelescopes (such as the Westerbork Array and the Very Large Array) at centimeter and decimeter wavelengths made possible observations of a few arcseconds resolution. Source sizes of 3 arcsec have commonly been observed at 15 GHz during flares (somewhat larger sizes are seen at lower frequencies, presumably from loops located higher in the atmosphere). Radio observations with high spatial resolution have been used to pinpoint the site of initial energy release in some flares, and such observations were the first to show the expansion and thermalization of high-temperature flare plasmas following the impulsive phase of the flare.

Further evidence for small-scale structure in the solar atmosphere comes from observations with high temporal resolution. Microwave spikes of millisecond duration are sometimes observed during flares, implying source sizes of the order of a few hundred kilometers. Even smaller structures are suggested by decimeter-wavelength observations of short-lived, narrow-band bursts. Direct observations of such structures at radio wavelengths may become possible during the next decade with the advent of the Very Long Baseline Array. More limited information might be obtained with existing very long baseline interferometers.

While high-angular and high-temporal resolution radio observations have yielded important new results, they are limited in one sense. All radio emission mechanisms are continuum in nature, and represent the signature of either thermal or energetic electrons. Therefore, radio observations do not yield information on the variety of ion species present in the solar atmosphere.

High-Angular-Resolution Optical Imaging from the Ground

There are a number of techniques that are currently yielding high-angular-resolution information at visible wavelengths on solar features. Adaptive optics, which use real-time measurement of atmospheric aberrations and phase correction of those aberrations, has demonstrated a capability for substantive image enhancement of small solar features (pores) and the surrounding granulation. Speckle interferometry and imaging have been applied to broadband sunspot and granulation data, yielding detail approaching the diffraction limit of a 1.5-m-aperture (0.07 arcsec). Aperture synthesis has shown promise of recovering enhanced imagery from a single short exposure, allowing very high time resolution.

As these and other techniques are improved and used on large ground-based telescopes, the requirements for future space-based systems will be greatly clarified. Of course, ground-based systems have the severe limitation of the atmospheric transmission window.

Long-Baseline Optical Interferometry

There has been recent, worldwide development of long-baseline (up to 1 km) interferometers. All such instruments have been designed for stellar astronomy. However, these or similar instruments could be applied to high-angular-resolution solar measurements. Special techniques that use field stops to isolate small features would be necessary to actively control the interferometric baseline, but these techniques are quite feasible. While no such solar observations have yet been proposed, such a system would have the potential of providing angular resolution approaching 1 milliarcsec at visible wavelengths. However, because of their sparse coverage of the (u,v) plane, such observations are more difficult to interpret than images proper, particularly for complex, low-contrast sources like the Sun.

SUMMARY AND CONCLUSIONS

There is now considerable evidence that all scales of structure on the Sun, as well as other astrophysically interesting objects, are strongly coupled to small-scale processes associated with intermittent magnetic fields and turbulent stresses. Understanding the physics of these dynamical structures and their interaction with their surroundings is essential for a proper description of large-scale structures (such as coronal active regions, flares, or the solar wind) and their effects on interplanetary space and the near-Earth environment.

The interplay between processes occurring on vastly different spatial scales is ubiquitous in astrophysics. Whether in accretion disks feeding black holes at the center of active galaxies or quasars, in the magnetospheres of neutron stars, or in the x-ray coronae now known to surround a wide range of stars, small-scale magnetohydrodynamic processes are thought to influence and sometimes control the behavior of the object.

In these astrophysical situations, observations using even the most advanced technology currently conceivable will not allow us to directly observe the controlling small-scale processes. Using the Sun, however, we can indeed imagine direct observations. The Sun is therefore a unique tool for advancing our understanding of a broad class of important astrophysical phenomena, if we can penetrate to the domain of underlying processes that often operate on spatial scales of 1 to 100 km.

An orderly progression of goals that could realize much of this promise would include:

1. Implementation of the High Resolution Solar Observatory on Spacelab, followed by the transfer of HRSO to the Advanced Solar Observatory on the Space Station, together with ultraviolet and x-ray solar instruments capable of 0.1-arcsec angular resolution.

2. Development of interferometric experiments at visible and ultraviolet wavelengths aimed at preliminary reconnaissance of solar features at angular scales much less than 0.1 arcsec.

3. Development of meter-class facilities exploiting the emerging multilayer coating technologies, designed to obtain resolution in the 0.01-arcsec regime at extreme ultraviolet or soft x-ray wavelengths. Ideally, several spacecraft would be located in near-Sun orbits to provide high flux, high-linear-resolution, and stereoscopic imaging.

4. Achievement of ultrahigh-resolution imaging in ultraviolet and visible wavelengths, using baselines of order 10 m.

Two concepts related to these goals are presented in separate recommendations of the task group: stereoscopic imaging using meter-class telescopes at four locations around the ecliptic, and an x-ray/ultraviolet/optical telescope in heliostationary orbit at 30 solar radii.

APPENDIX: LIST OF PARTICIPANTS

S. Antiochos, Naval Research Laboratory
R. G. Athay, High Altitude Observatory, NCAR
J. Beckers, Advanced Development Program, NOAO
R. Bonnet, European Space Agency
G. Brueckner, Naval Research Laboratory
R. Catura, Lockheed Palo Alto Research Laboratory
L. Cram, Commonwealth Scientific and Industrial Research Organization
L. Dame, Laboratoire de Physique Stellaire et Planetaire
J. -P. Delaboudiniere, Laboratoire de Physique Stellaire et Planetaire
G. Doschek, Naval Research Laboratory
G. Epstein, NASA, Goddard Space Flight Center
T. Gergely, NASA, Headquarters
M. Giampapa, National Solar Observatory, NOAO
L. Golub, Harvard-Smithsonian Center for Astrophysics
J. Harvey, National Solar Observatory, NOAO
J. Heyvaerts, Observatoire de Meudon
T. Holzer, High Altitude Observatory, NCAR
R. Howard, National Solar Observatory, NOAO
R. Keski-Kuha, NASA, Goddard Space Flight Center and University of Maryland
J. Leibacher, National Solar Observatory, NOAO
B. Lites, High Altitude Observatory, NCAR

R. MacQueen, High Altitude Observatory, NCAR
P. Nisenson, Harvard-Smithsonian Center for Astrophysics
R. Noyes, Harvard-Smithsonian Center for Astrophysics
D. Rabin, National Solar Observatory, NOAO
F. Roddier, Advanced Development Program, NOAO
R. Rosner, Harvard-Smithsonian Center for Astrophysics
L. Schmutz, Adaptive Optics, Inc.
E. Shoub, University of Colorado
D. Spicer, NASA, Goddard Space Flight Center
A. Title, Lockheed Palo Alto Research Laboratory
J. Toomre, University of Colorado
J. Underwood, Lawrence Berkeley Laboratory